Power Plant Technology

Dr.P. Balashanmugam

G. Balasubramanian

Published by

Power Plant Technology

ISBN 978-93-86638-42-7

Authors

Dr.P. Balashanmugam

G. Balasubramanian

Bonfring
309, 2nd Floor, 5th Street Extension, Gandhipuram,
Coimbatore-641 012.
Tamilnadu, India.
E-mail: info@bonfring.org
Website: www.bonfring.org
Phone: 0422-4213231

Preface

We are immensely pleased to bring out this book *'Power Plant Technology'* for the Mechanical Engineering Students. From the teaching experience of last 14 years of undergraduate classes, it appeared to author that there was a need for a concise and practical book on *Power Plant Technology*. The treatment of the book begins with the fundamentals and develops in a way that allows the practical man as well as the student to expand his knowledge progressively.

Chapter 1 deals with the thermodynamic cycles for power plants and constituents of modern thermal power plants.

Chapter 2 deals with station layout, coal handling system and factors affecting site layout.

Chapter 3 deals with use of coal in electricity generation and Inventory of Coal Reserves of India.

Chapter 4 has an in depth dealing of steam boiler and its working cycles. More specifically, it deals with Fluidized bed boilers, burners and controls.

Chapter 5 deals with the boiler control system flame monitoring and feed water control system.

Chapter 6 deals with the preparation for commissioning procedure, testing of boiler and efficiency evaluation.

Chapter 7 deals with the types of turbines and turbine governing system.

Chapter 8 deals with the methods of governing and programmable logic controller.

Chapter 9 has an in depth dealing of turbine installation and operation, starting up of turbine and turbine shut down procedure.

Chapter 10 deals with the technical options for co-generation and the development trend of CHP plants.

Chapter 11 deals with the Emergency cases requiring the shutdown of the boiler, Shut down of the unit requiring the use of vacuum breaker, Shut down of the turbine without use as vacuum breaker and Emergencies in Generator and electrical Equipment.

Chapter 12 deals with Emissions from Thermal Power Plants, Emissions from coal usage, Emissions from Coal Fired Thermal Power Plants in India and Energy Activities.

Chapter 13 Thermal Power Plant Economics, Requirement of thermal power plant design, Reduction in operation and maintenance costs, Cost of electrical Energy and Selection of type of generation.

Acknowledgement

No book is ever written without the assistance of a great many people. Here we would like to acknowledge a few of them. First and foremost, We extremely grateful to Professor **AR.Ramanathan**, formerly Head, Department of Mechanical Engineering, Annamalai University for his valuable guidance and suggestions.

We would like to express my gratitude to **Dr.C.G. Saravanan, M.E., Ph.D**, Head of the department of Mechanical Engineering, Annamalai University for his kind cooperation and encouragement which helped in the completion of this work.

We express my thanks to professor **Dr.C. Antony Jeyasehar**, Dean, Faculty of Engineering and Technology for his constant support and encouragement.

We would also like to thank all the faculty members of the Mechanical department who have co-operated and encouraged during the study course.

We would also like to thank all the staff (technical and non-technical) and librarians of Annamalai University, Annamalai Nagar who have directly or indirectly helped during the course of my study.

We deeply indebted to my respected teachers and other members of the English department for their invaluable help in preparing this book.

I humbly extend my thanks to all concerned persons who co-operated with me in this regard.

We would like to thank the many people who have made this book possible

Finally, we would like to thank my family & friends for their constant love and support and for providing me with the opportunity and the encouragement to pursue my goals.

Dr.P. Balashanmugam
G. Balasubramanian

Author's Profile

Dr.P. Balashanmugam received his B.E. in Mechanical Engineering from Annamalai University, in India, in 1998, his M.E. in Energy Engineering and Management from Annamalai University, India in 2007 and his Ph.D., in Automobile emissions and air pollution modelling from Annamalai University, India in 2014. He is currently Assistant Professor in the Department of Mechanical Engineering, Annamalai University, Chidambaram, Tamilnadu, India. His research interests are in the areas Automobile emissions/pollutions, Air pollution monitoring and modelling and renewable energy sources (solar energy and applications). He has published more than 50 papers in national, international conferences and journals. He is a life member of the Combustion Institute.

G. Balasubramanian received his B.E. in Mechanical Engineering from Annamalai University, in India, in 2005, his M.E. in Thermal Engineering from Annamalai University, India in 2010 and. He is currently Assistant Professor in the Department of Mechanical Engineering, Annamalai University, Chidambaram, Tamilnadu, India. His research interests are in the areas Automobile emissions/ pollutions, Air pollution monitoring and modelling and renewable energy sources (solar energy and applications). He has published more than 30 papers in national, international conferences and journals.

1. Principles of Modern Thermal Power Station

1.1. Introduction

Basically, a thermal power station provides a mechanism to transform energy. The chemical energy that is released from hydrocarbons in fossil fuels when reaction with oxygen takes place at a high temperature (which process is known as combustion) is the input to the power plant and the electrical energy delivered by the electrical generation is the output. In a conventional power plant the energy supply (input) is first converted to an electrical generator.

Thus the energy conversions involved are:

Heat energy → Mechanical work → Electrical energy

The first energy conversion takes place in what is called a boiler or steam generator, the second in what is called a turbine and then last conversion take place in the electric generator. As is the case in many processes which involve an energy conversion, it is not possible to convert the first energy form to the ultimate form in the thermal plant without a third undesirable conversion taking place viz. Heat rejection and energy transfer in the exhaust (chimney).

1.2. Thermodynamic Cycles for Power Plants

It is known that the laws of the thermodynamics determine the manner in which processes involving heat and work interactions may proceed. A thermodynamic cycle is a series of processes combined in such a way that the thermodynamics states at which the working fluid exists are repeated periodically.

In a conventional electric generating station the fluid is cycled through a sequence of processes in a closed loop designed to produce the maximum generation of electrical power from the fuel consumed consistent with the plant economics.

The basis of the modern steam power plant is the Rankin cycle, which is a modification of the Carnot cycle. This cycle, which can also be called an ideal saturated steam cycle and enthalpy-entropy diagram together with a schematic diagram of the power plant, is shown in Fig 1.1. Unlike in the case of the Carnot cycle, in the Ranking cycle the condensation process accompanying the heat rejection process continues until the saturated liquid state is reached. And a pump replaces the two-phase compressor found in the former cycle.

Rankine Cycle Schematic

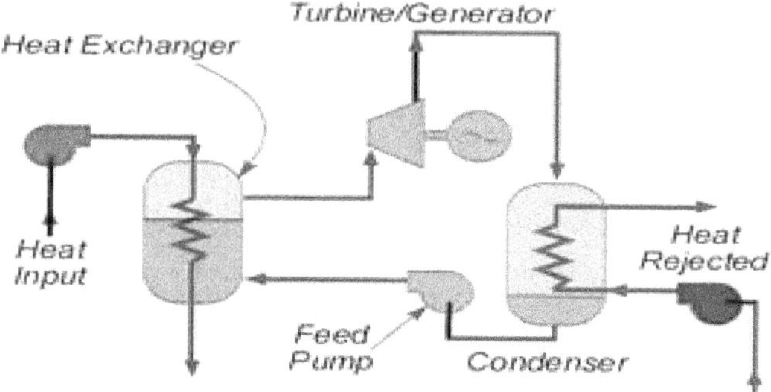

Fig. 1.1: Rankine Cycle

The Ranking Cycle is comprised of the following processes:

1-2: Heat supply at constant temperature and pressure

2-3: Isentropic expansion with work output

3-4: Heat rejection at constant temperature and pressure carried out until condensation is complete.

4-5: Isentropic pumping process to raise the liquid pressure to that of the boiler

5-1: Heat supply at constant pressure, but not constant temperature carried out until all liquid is in a saturated state.

Economic advantage can be obtained by modifying the cycle so that the maximum temperature in the cycle is increased. This method to employ the superheat cycle is illustrated in Fig.1.2. In this cycle there is an additional heating process $2\text{-}2^1$ at constant pressure.

This results in increase of thermal efficiency of the cycle. The heat added is the area under the irregular line $5\text{-}1\text{-}2\text{-}2^1$. The heat rejected (in the condenser) is the area under the line 3-4. The isentropic expansion in the prime-mover (turbine) is the vertical line $2^1\text{-}3$ and the isentropic compression in the (feed pump) is the vertical line 4–5. Usually the compression phase is so small as to be negligible and the prints 4 and 5 can, therefore, be considered as a single point, saturated liquid at exhaust pressure. The net work output of the Rankine Cycle is then given by

$$\text{Net work} = h_2' - h_3$$

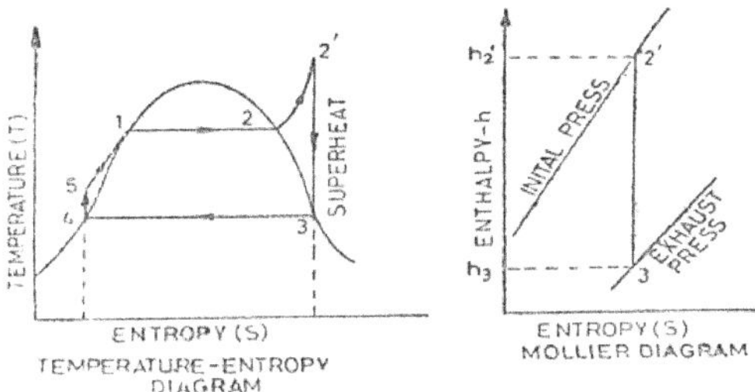

Fig. 1.2: Modified Rankine Cycle

The heat supplied to make steam is $h_2' - h_4$

The efficiency (approximate) of Rankin cycle

$$= \frac{h_2' - h_3}{h_2' - h_4}$$

The heat rate, which is the heat supplied per unit of power output, is generally preferred to thermal efficiency because it is more directly applicable to fuel performance. The relevant relationship is:

$$\text{Heat rate} = \frac{860}{\text{Thermal efficiency}}, \text{Kcal per KWh}$$

1.3. Reheat Cycle

The thermal efficiency is further improved be reheating the steam after partial expansion followed by continued expansion to the final exhaust pressure. This cycle shown in Fig. 1.3 has found wide acceptance in large power stations.

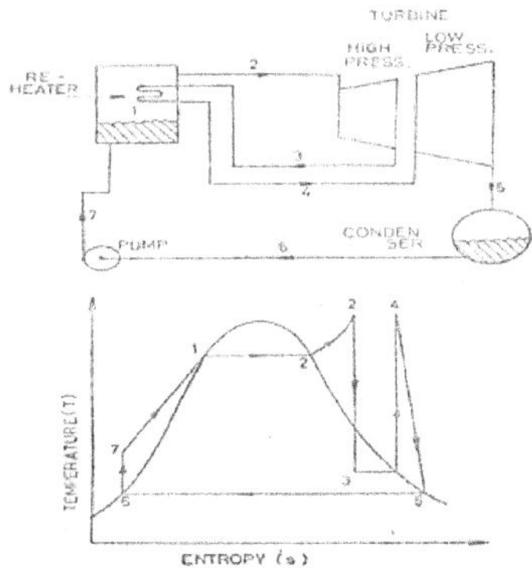

Fig. 1.3: Reheat Cycle

Fig. 1.3 shows the effect on ideal cycle thermal efficiency of reheating at different initial throttle pressures. Throttle and reheat temperature remain constant at 538ºc (1000ºF) at different pressures. It will be noted that the efficiency in maximum for a reheat pressure that is 20 per cent of the initial throttle pressure. It will also be noted that the cycle efficiency is improved to increase in initial throttle pressure.

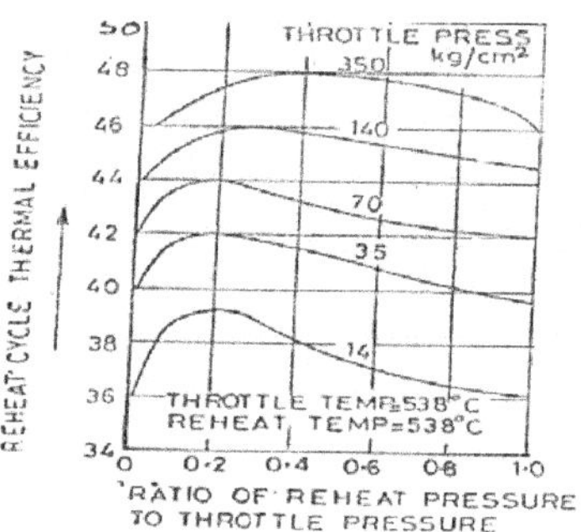

Fig. 1.4: Relation between Reheat and Pressure Thermal Efficiency

1.4. Regenerative Cycle

The practical steam power plant uses a regenerative cycle shown in Fig. 1.5. This cycle is attempting to impart reversibility in the Rankine cycle and consequently increase its efficiency. In the cycle shown in Fig. 1.1, the principal irreversible process was the mixing of cold condensate with the mixture of saturated steam and water in the boiler. The regeneration aims at reducing this irreversibility by heating the condensate near to the saturation temperature through the utilization of some otherwise waste heat source. The steam for regenerative heating is extracted or "bled" from an intermediate stage of steam expansion in the turbine such that maximum work would have already been obtained from the live steam and its latent heat of the balance steam is lost to the circulating water. Thus the heat in the extracted steam is retained in the system whilst the latent heat of the balance steam is lost to the circulating water. In other words a certain percentage of the total steam performs 100 per cent.

Work in the system, thereby raising the overall efficiency of the plant. Incidentally, the surface of the condenser can be reduced marginally as it would now be required to handle the lesser quantity of steam. Multistage regenerative heating is normally used in keeping with the thermodynamic desirability of equal temperature rises in all stages. The effect of feed temperature and number of staged of heating on the gain in thermal efficiency with ideal regeneration cycle is shown in Fig. 1.6.

1.5. Constituents of Modern Thermal Power Plant

It has been from article 1.2 that a thermal power plant will have certain basic equipment arranged in a particular sequence so that energy transfer/transformation takes place in an optimum way. It could be even stated that these components are inescapable. The basic elements of power plants can be stated as fuel, air and water. The actual system of power generation, however, is quite complex and these include fuel and ash handling; transport of air and combustion products; feed water, steam, condensing plant, control functions, and transmission of electric power output. These systems are very diverse in nature and belong to different disciplines of science and technology. The power plant is a complex of large dimensions involving huge amounts of resources, both material and human. Consequently the design, construction, operation and maintenance of thermal power plants involve effective management of resources as well as personnel.

The following major system of a power plant has already been identified.

- Boiler
- Turbine
- Generator
- Condenser
- Regenerative system

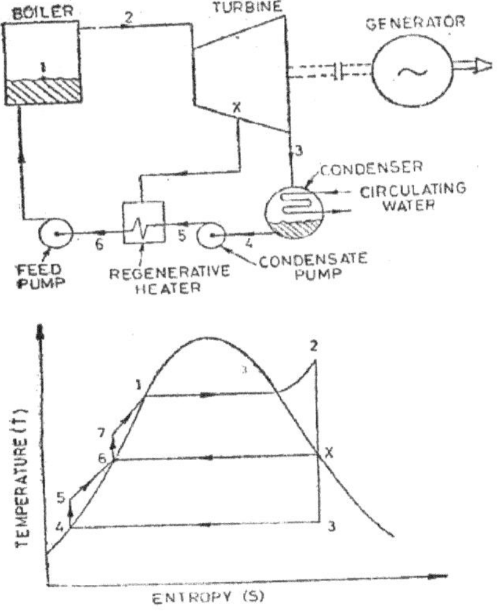

Fig. 1.5: Regenerative Cycle

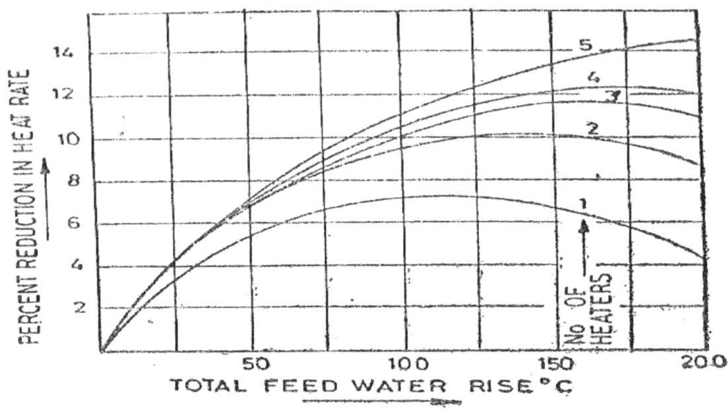

Fig. 1.6: Gain in Thermal Efficiency

1.6. Boiler

A boiler is essentially a container into which water can be fed and, by the application of heat, evaporated continuously into steam. Such an arrangement was, however, very inefficient. It was necessary that more water should be in contact with heat. One way of achieving this was to direct the hot products of combustion through tubes submerged in water. Such an arrangement is called a fire tube boiler. Another arrangement was to arrange one or more drum interconnected by means of tubes so arranged that tubes is transferred to the mixture of water and steam in the tubes. This arrangement is called a water tube boiler. In the water tube boiler the pressure parts are subdivided and hence large capacities are possible. This form of boiler design is, thus, most commonly used in power station application. Although the first water tube boiler was introduced in 18th century, it has undergone significant changes in the past 40 years.

The first input having been identified as fuel; let us take a look at the constituents of the fuel firing system. The fuel firing system involves the fuel, the method of placing it in the boiler and the arrangement for burning it properly. The process of burning the fuel is a continuous one, meaning that the fuel has to be fed to the furnace continuously at a predetermined rate. The rate at which the fuel is fed would depend naturally on the rate at which steam is to be generated for a given fuel and boiler arrangement. When coal is used as fuel, the residue of combustion ash is also required to be removed continuously so as to make room for fresh coal to come in. Fig.1.7 shows a fuel burning system in its very simple form.

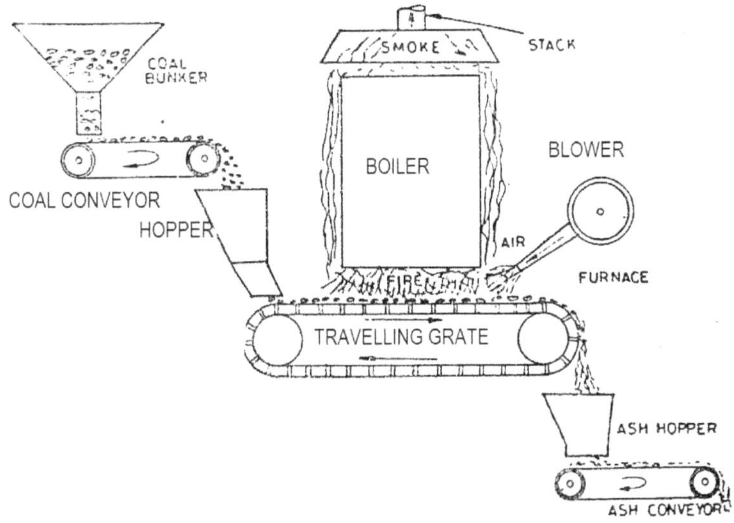

Fig. 1.7: Constituents of Fuel Burning System

Combustion is a form of oxidation process in which oxygen combines rapidly with fuel and substantial amounts of heat are liberated. Coal contains two basic fuel elements, carbons and hydrogen, in combinations called hydrocarbons. Sulphur and some other elements also burn to give off heat but the contribution of these to the total heat is nominal. The way carbon burns have been explained as the oxygen penetrates the carbon surface to break away atoms, which hook up with oxygen in a loose unstable carbon- oxygen compound. Depending on the temperature and other conditions, this compound breaks up into carbon dioxide and carbon monoxide. If there is an excess of oxygen, carbon monoxide is oxidized to carbon dioxide and if carbon is in excess, dioxide is reduced to monoxide.

Each element or compound requires a certain amount of oxygen for complete combustion. It is possible to compute exactly how much oxygen is required as also how much air is needed. By mixing the quantity of air, thus calculated with the fuel a perfect or theoretical mixture is obtained. The relation between the fuel and air called the fuel air ratio. Any fuel burning system is required to carry out the following tasks:

1. Prepare the fuel and air.
2. Convert fuel into elementary fuels.
3. Bring these elementary fuels and air together in the right ratio and at the proper temperature for ignition and combustion.

4. Transfer the heat in the product of combustion in the boiler (steam generating parts) while retaining enough heat is the combustion zone to maintain volatilization and ignition.

The fuel is prepared in pulverizes in which the size of coal is reduced to small particles. It will be appreciated that small particles show more area to heat. The pulverize is not an integral part of the combustion chamber and hence a means of transportation is required to be introduced. Directing the primary air through the pulverizer enroute the burner carries this out. The primary air ensures fast motion between particles and air to strip away the microscopic gas envelops that tend to insulate the particles against the heat.

In the present day system such air is called upon to drying the coal of its surface moisture and classifying (size of pulverizing) also. The drying function of air makes it necessary so much so the primary air is necessary to be preheated. In addition to the primary air, air for combustion called secondary air is also introduced into the furnace along with the coal-primary air and the secondary air is carried out by utilizing the waste heat in the flue gases.

The air and fuel supply to the furnace through burners. Burners are required to provide firstly stable ignition, secondly effective control of flame shape and travel and lastly through and complete mixing of fuel and air.

When the minute particles of the pulverized coal enter the furnace these are exposed to the prevailing radiant heat. The temperature of the particles rises and the volatile matter distills off. The gases (carbon-oxygen compounds) intimately mix with the primary air and burn. The percentage of primary air required is approximately equal to the percentage of volatiles. The hydrocarbons in the volatiles first ignite and intern heat the carbon content to incandescence. The secondary air that is introduced around the burner carries away the incandescent carbon particles and gradually burns with them over the length of the flame. Thus the flame would have to be several feet long and proportional to the time of combustion. For effecting combustion, the velocity of the primary air coal mixture must be greater than the speed of flame propagation. However, at the beginning, for initiating the process of ignition, auxiliary arrangement is made in the burner. A separate sub - system using either oil or gas or both with the spark for ignition provided electrically is normally provided. Even with the best arrangement of burning system certain amount of unburned combustibles find it difficult to find oxygen and consequently these escape through the flue gases. The combustibles escaping can be brought down by providing a high degree of turbulence. The percentage of excess air and ash content in the fuel. The percentage excess air will have to be increased with an increase in the ash content. A balance is required to be struck, as too high an excess air will result in loss of sensible heat in the flue gas.

It has been seen from the above that the constituents of the fuel firing system are:

a)   Raw coal hopper or bunker to store coal.

b)   Raw coal feeder to carry desired quantity of coal from bunker to pulverisers.

c)   Pulveriser.

d)   Burner and wind box.

e)   Furnace.

f)   Primary air fan (P.A.Fan)

g)   Air pre heater

h)   Secondary air fan or forced draught fan (F.D.Fan).

This system can schematically be shown as in Fig.1.8.

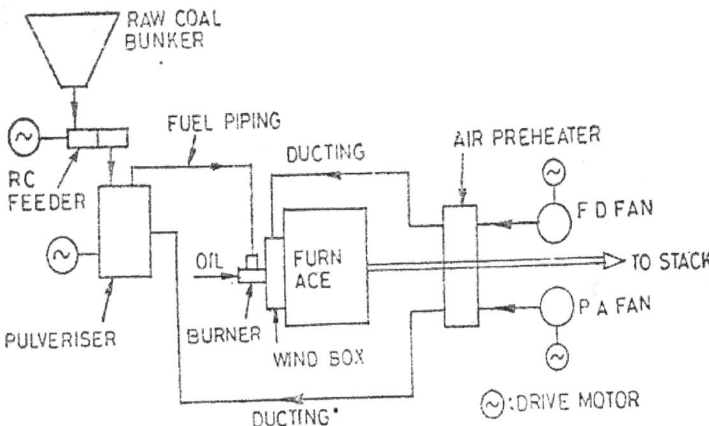

Fig. 1.8: Fuel Burning System

It may be noted that a particular steam-generating unit may have more than one pulveriser depending on the fuel requirement. Similarly a plant may be equipped with more than one F.D fan and P.A. fan. Burners also will be numerous and depending upon the arrangement of burners there may be more one-wind box. We have seen how a large amount of heat is liberated in the furnace. The temperature attained is of the order of 1200 to 1400°c whereas the metallurgical considerations on the steam raising parts impose a maximum limit of 540ºc as the maximum attainable temperature of the steam.

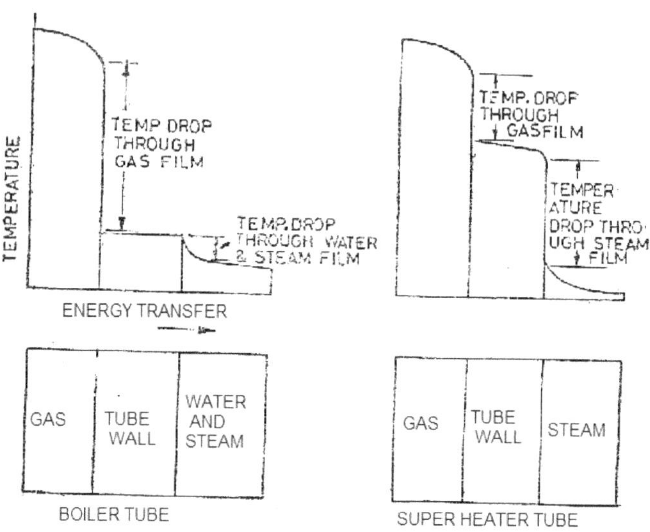

Fig. 1.9: Heat Transfer in Boiler

It is thus necessary to remove effectively the large amount of heat liberated and to handle the high temperatures resulting from radiation of incandescent fuel particles. In order to obtain more efficient and economical steam generation, the modern boilers are provided with heat absorbing surface virtually enclosing the furnace. As seen in Fig. 1.9 heat transfer is maximum in the case of water walls. They receive heat by conduction and convection from hot combustion products. They also receive heat by radiation from the incandescent fuel particles. In modern boilers nearly the water wall makes up 50 percent of the total heating surface. Water walls are connected to the boiler drums through down comers and headers are suitably arranged to which the water wall tube panels are connected. The headers in turn are connected back to the drum.

Saturated steam from the drum is further heated in a super heater located in the upper portion of the furnace space where the temperatures are still high and in some cases radiation effect also is utilized for super heaters. In a modern boiler the super heaters may account for mover 16 per cent of heat absorbed. Beyond the super heaters in the path of the flue gases the temperature would have reached a level where any further heat recovery of recovery by steam raising surface would not be possible. The lower heat level portions of the cycle are located in the zone, which is called the convection pass. Since feed water temperature is below saturated steam temperature, it is possible to recover some heat from the flue gas by locating what is called an economizer in this path. The economizer absorbs about 6 percent of total heat absorbed.

A further heat recovery is made possible by locating the air heater through which heat is transferred to a still lower temperature fluid the air about 10 percent of heat is absorbed in air heater.

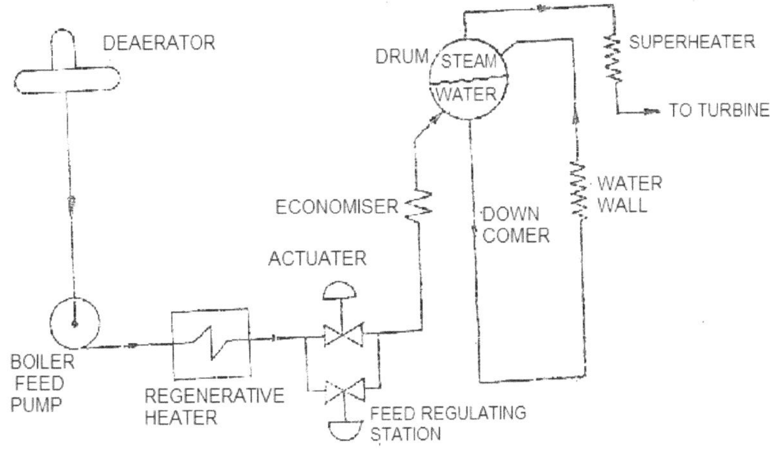

Fig. 1.10: Water and Steam Circuit

The boiler drum is located at an elevated position so that all boiler tubes connected to it are filled with water. The drum is at a high pressure and consequently any fresh water supply to the drum would have to be pumped against the pressure and then static head. Boiler feed pumps carry out this job. The supply of water should be closely regulated so as to match the stem load output and for this purpose a feed water regulation station where control valves are located in the feed water line is used. The feed regulation station is located near the inlet to economizer enroot to the boiler drum. Fig.1.10 schematically represents the water and steam circuit of a power station boiler.

2. SITE AND STATION LAYOUT

2.1. Introduction

The various requirements of a thermal power station and how they influence the selection of a site were dealt. Consideration of these factors would only lead to a general location of the site, with reference to the source of coal, source of water etc. Specific and exact location of the site determining the relative location of the plant constituents i.e. site layout will have to be decided after further critical examination of techno-economic analysis. The objective of site layout selection should be to fine an optimum arrangement, which results in minimum overall cost consistent with ease of erection, ease of operation and good appearance within the limitations of the available site.

2.2. Ultimate Capacity

In developing the site layout for a thermal power station, the capacity of the station for immediate installation as well as the ultimate installed capacity should be kept in mind. Let us examine how the ultimate installed capacity affects certain facilities and how this is required to be catered to.

2.3. Cooling Water System

In the case of direct system, it will be necessary to examine the effect of the ultimate capacity on the size of the intake system. In case of an installation with 5X200MW units planned in the stage I and 2X500MW under stage II, it would be necessary that about 13 no's of CW pumps are installed under each stage. Locating all the 26 pumps in one pump house might make it very unwieldy apart from the fact that the arrangement to feed large amounts of power might pose problems. One of the consequential effects would be an increase in the capacity of the station transformer, it is decided to feed the CW pump house from the station supply system and in case it is assigned to the unit, auxiliary supply system, the number of feeders that would have to be handled will be necessary that the CW pump house is planned in two sections each located to suit the convenience of the water conducting system and power supply arrangement. The decision will, however, have to be taken after a thorough examination of the techno economics. The route that the intake duct/culvert for the two stages will take might dictate the relative location of the two stages of the project. The closed circuit cooling system will afford some degree of flexibility on this account, as the water required to be handled is not much.

2.4. Coal Handling System

While locating the coal handling plant it will be necessary to ensure adequate area is made available for the reserve coal storage for ultimate requirements. The location of coal yard should be decided keeping in view the wind direction for, however much care is taken to ensure no dust rises, it can be a source of dust to the main plant as also the residential area(staff colony). Sufficient space should be left in the marshalling yard for all tracks that would be required for the ultimate capacity of the station.

2.5. Switchyard

While locating the switchyard the ultimate level of voltage in keeping with the quantum of power required to the evacuated under ultimate generation conditions should be decided initially itself and adequate space for the right level voltage equipment should be provided. There have been instances where provision was made initially for 200KV switchyard and on subsequent addition in generating capacity 400KV levels was required and it has not been very easy to find a convenient location or space for this higher voltage equipment. Care is required to be taken while providing corridors for the transmission lines that take off from the switchyard. Crossing of lines of different voltage level should be avoided. It would be advisable to avoid having the switchyard and coal yard on the same side of the plant so as to obviate coal dust pollution of the switchyard equipment and insulators.

2.6. Colony and Amenities

Staff colony and amenities that go with any decent township should be planned with the ultimate capacity in mind. Not always that this important aspect gets its deserved consideration.

2.7. Other Common System

Ash disposal area, fuel oil storage space, hydrogen generating plant, clarified water system, administrative building, are other system and facilities that are required to be conceived with the ultimate capacity in mind. Some example of lack of perspective planning can be found where generating units have been added as an afterthought. This had led to cramped and dissimilar layouts, needed certain make shift arrangement and, to some extent, had compromised on the requirement of good engineering practice. For obvious reasons such bad examples are found in urban-based power stations.

2.8. Factors Affecting Site Layout

There cannot be any standard layout of power station even with similar ratings. The site layout will have to be evolved to suit the conditions particular to a given site.

The main factors that influence site layout of any thermal power station are given below:

1. Foundation
2. Proximity to coal mines
3. Ground levels
4. Circulating water system
5. Coal and fuel oil handling systems
6. Ash disposal system
7. Road and rail access
8. Auxiliary buildings
9. Construction inputs
10. Evacuation of power
11. Colony and amenities.

2.8.1. Foundation

The power station structures impose load up to 60 tonnes per sq.metre on the sub-soil. Not all soils will have this bearing strength. Piled foundations will be required in case the bearing strength is falling short of the requirements. This will add to the cost of the civil works. For soils of medium strength mat foundation will be required, the cheapest foundation, i.e. isolated column type being suitable for rocky soils. Thus the cost of civil works will vary depending upon the soil condition.

The bearing strength of the soil is determined by driving a number of boreholes and analyzing/testing the samples. It is possible that in a given site there are certain spots where rock-bearing soil is obtained. By locating heavy structure like a chimney, wagon tippler, crusher house, boiler house, etc. in such spots considerable economy might be possible. Where the land is undulating, building of retaining wall swill be required thus adding to costs.

2.8.2. Proximity to Coal Mines

In case of pit-head station, it will be necessary that sufficient precaution is taken so that mining would not be undertaken within a closely defined volume of ground near the power station structures.

2.8.3. Ground Levels

The variation in levels of the ground will have its influences on the following:

1. Pumping costs of circulating water
2. The cost of civil works in levelling
3. Requirement of flood protection measures
4. The extra time taken in completing civil works.

It will be necessary that economics are worked out for alternative levels before and optimum level is adopted. It is not possible to have the same level throughout the plant site as this might involve a large amount of excavation work.

2.8.4. C W System

Layout of circuit water system requires careful consideration. Availability or otherwise of water for adoption of direct cooling will greatly affect the amount of land required for this system. The area required for the cooling towers will be 10 percent of the total area of the plant.

2.8.5. Coal and Fuel Oil System

Almost all power stations in operation in the country other than mine mouth station receive coal by rail. One exception to this is the recently constructed Tuticorin power station in the southern parts of the country, which is scheduled to receive seaborne coal. In case of pithead stations some receive coal by belt conveyors and some by ropeways. Neyveli Thermal Power Station in Tamilnadu state receives lignite by means of belt conveyor whereas Bokaro thermal power stations receives coal by rope way.

Coal fired stations require fuel oil and, in some cases, light oil also. Fuel oil is used for light–up and at low loads. Where light oil is used, it is employed to establish ignition energy in the furnace. In some cases it may be necessary to continue feeding fuel oil at all loads for maintaining flame stability. Arrangements and facilities for receiving, preparing (heating) and pumping are made/provided. The oils are received on rail in tankers and stored in tanks. As the fuel oils are a fire hazard it is necessary that it is away from any other plat facilities.

2.8.6. Ash Disposal System

A dry system for ash disposal is inadequate for large stations in view of vast quantity required to be handled. Wet disposal is thus invariably used in large plants. It is estimated that about 30 sq.km of area is required for the life time of a 2000 MW plant burning 40 percent ash content coal if the stacking height is 3 meters. Load pipelines and high pumping costs will have to be avoided.

2.8.7. Evacuation of Power

About 20 percent of the total plant area will be required for the switchyard. Level ground with proper drainage is required. As already indicated, it is advisable to keep it away from the coal storage area to avoid dust pollution. On the output side, it has to be ensured that sufficient corridors are available for routing the transmission lines. For easy connection between the unit step-up transformer and the switchyard, the latter is located next to the transformer yard, which it will be next to the turbine hall so that the generator terminals could be connected to the step-up transformer by means of a short bus duct. Switchyard out-door type are used in this country.

2.8.8. Other Consideration

Access by rail and road are necessary in view of the heavy equipment required to be transported to the site. It is preferable it the power station is planned to be self-sufficient in so far as testing and maintenance resources are concerned. Laboratories covering chemical, electrical, electronics, instrumentation disciplines should be provided fully equipped and is dust-free. Power stations, especially large ones are located away from major towns. A residential quarter, township with all amenities is thus required to be provided. Educational centers, banks, recreation facilities, parks, shopping centres shall cover essential amenities.

2.9. Station Layout

A modern thermal power station contains several major sub-systems all having their individual peculiarities and each one of them has its own importance in the plant performance. The boiler–turbine generator (B.T.G) nit takes coal, feed water for its inputs and gives out electric power at the output. In this process, the boiler gives out a lot of ash and flue gas and the turbine needs large amounts of cooling water. From economical point it would be desirable to have all the sub-systems as close to the B.T.G unit as possible. Coal handling plant, C.W unit system, fuel oil plant, and ash disposal system would need to be at some distance to the B.T.G unit. Further, their relative location with respect to the B.T.G unit also cannot follow a set standard pattern for all stations, but will depend on the situation particular to a given station. Thus the endeavour could be to obtain an optimum distance between the B.T.G unit and the sub-system consistent with the overall economy and convenience of plant operation and control.

It follows from the above that the plant equipment in a station can be grouped under two distinct group's viz. main plant and auxiliary plants. The main plant consisting mainly of boiler, turbine and generator will cover their respective auxiliaries also. The auxiliary plants, let us call these station auxiliaries also, differentiate from the auxiliary system of the B.T.G nit, will cover the balance plant equipment in the station. The equipment under these two groups can be written as follows.

2.9.1. Main Plant

1. Boiler unit
2. Mill system
3. Draught system
4. Precipitator
5. Stack
6. Turbine
7. Condensate system, including deaerator
8. Lube oil system
9. Feed water system
10. Transformer
11. Generators
    A. Unit auxiliary supply system
    B. Unit control board

2.9.2. Auxiliary Plant

1. Coal handling system
2. C.W. System
3. Ash handling system
4. Fuel oil handling system
5. Water treatment plant
6. Air conditioning plant
7. Compressed air system
8. Instrument air system
9. Fire fighting system
10. Clarified water system
11. Chlorination water system
12. Hydrogen generation plant

13. Start-up boiler

14. Switchyard

15. Station auxiliary supply system

16. Station control board

2.9.3. *Electrical Equipment*

The major electrical equipment that requires consideration in the layout of main plant are:

1. Turbo-generator
2. Main generator transformer
3. Unit auxiliary transformer
4. Station auxiliary transformer
5. Distribution transformers for LV supply
6. HV switchgear
7. LV switch gear and motor control centers
8. Cable system
9. Station, D.C. system
10. Bus ducts

2.10. General Design of Power Plants

In the design of a power generating station, consideration must be given to the arrangement of main auxiliary machinery. The various important factors, which are considered for the economic design of power plant, are volume and area required per KW capacity of plant, reliability of operation, ease of control, convenience of access to all important equipments and station cost per KW capacity.

The following of the factor for plant selection:

1. Volume and area requirement
2. Architectural features
3. Inline arrangement of units
4. Units perpendicular to each other or parallel division
5. One room one floor station
6. Unit plan station
7. Outdoor type stem electricity station
8. Lighting of steam power plants.

2.11. Volume and area Requirements

2.11.1. Volume and Area Requirements

Less volume as well as less area requirement is an essential feature of a good design of power plant. Less area requirement reduces the consideration cost of the plant. Some typical figures on the square meter of the floor area and cubic meter per KW of installed capacity for different components of a power plant as an average are shown in table 2.1.

Table 2.1: Typical Figures on the Square Meter of the Floor Area and Cubic Meter per KW of Installed Capacity for Different Components of a Power Plant

(Capacity between 100 to 250MW)

Station equipment	Boiler room	Turbine room	Switch house	Office	Intake works	Total
Floor area in m²per KW	0.026	0.015	0.003	0.003	0.0025	0.0495
Volume in m²per KW	0.8	0.5	0.04	0.1	0.06	1.5

2.11.2. Reliability

The reliability of operation is another important of design intimately associates with the cost of the plant. Many times, it has been proved by experience that the low cost plant is more expensive in operation than high cost plant due to non-reliability of operation. The use of inferior equipment or omission of certain apparatus necessary for precaution control always results in unsatisfactory operation of the plant. This is due to high outage of plant or high maintenance.

2.11.3. Ease of Control and Access to Equipments

The control system must be highly efficient in operation. This is very essential to avoid the total shutdown or breakage of the power plant under emergency conditions. Easy access to all equipments absolutely necessary for easy repair and maintenance cost.

2.11.4. Station Cost

The total station cost includes the cost of equipments, building and land. The total cost of the thermal plant lies between 1500 to 2500 rupees per KW of installed capacity. The construction cost of the plant can be reduced with careful check during the progress of the plant design. The net generating cost depends upon the station cost and operating costs. Therefore, the generating cost depends to a great extent upon the degree of refinement attempted in the reduction of heat rate and upon initial pressure and temperature of the steam used for power generation.

2.11.5. Architectural Features

In the design of modern power plant buildings, enough attention has been given to the architecture of the buildings. It is proved fact that better architecture of the buildings provides a psychologically pleasant atmosphere for the workers. The size, arrangement and shape of the building are principally affected by the type and size of the power plant. A very careful thought should be given in the design of the ventilation system of then a power plant without disturbing the look of the building due to the dust arrangement carrying the conditioned air. The roofs are generally constructed flat, with a covering of built-up tar and gravel composition. The roof deck is carried in trusses on reinforced concrete, as it is fire-resistant.

2.11.6. Inline Arrangement of Units

Majority of the modern power plants are provided with the division wall between the turbine room and boiler room. A typical arrangement of turbine and boilers being in line is shown in fig 2.1.

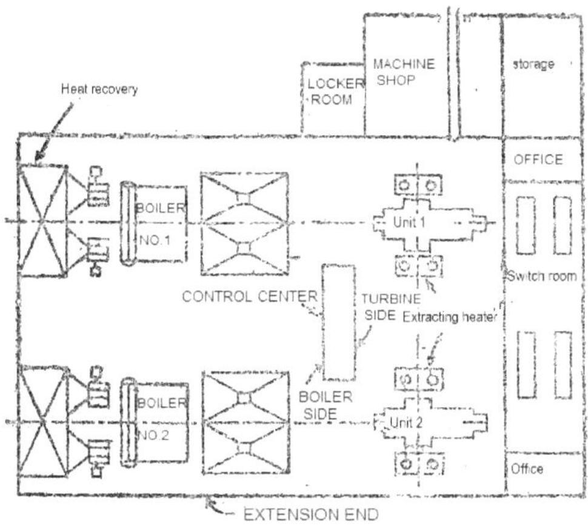

Fig. 2.1: Inline Arrangement of Units

The development of boiler room and the arrangement of prime mover depend on the number of boilers used per turbine. Now a day's boiler of high capacity is developed, therefore, the number of boilers used per prime mover becomes a matter of reliability. Some design engineers prefer three boilers for two turbines, one being a spare. But in large capacity plants, one boiler per turbine is economically justified if reliable electrical interconnections are available.

2.11.7. Unit Perpendicular to Each Other or Parallel to the Division

In this arrangement, the turbine units are set much closer compared with the previous arrangement. The condensers under the turbine have their longest dimensions, parallel to the turbine shafts to make close spacing possible. With this type of condenser arrangement, careful construction is necessary in order not be to block the space for turbine extraction lines, generator leads and air coolers. This arrangement is more flexible; therefore, turbine units of greater length can be installed in future without increasing the width of the turbine room. In old power plants of this arrangement, cross compounding of the turbine is used. The reason for adopting the cross compounding is the division of steam flow to two or more low pressure units and preservation of the initial turbine room width. During those days, a suitable large single shaft generator could not be constructed and, therefore, cross compounding was invariably used. Presently, in modern power plants, it is a matter of heat economically whether future machines will be of this type since a tandem compound steam turbine driving one generator can be built in sufficient large size to satisfy practically any load increase.

2.11.8. One-Room One-floor Station

The arrangement of this system is shown in fig 2.2. This new type of arrangement is developed to achieve maximum economy of design. In this arrangement, the division wall between turbine room and boiler room is totally omitted. This arrangement has been used successfully for several small stations.

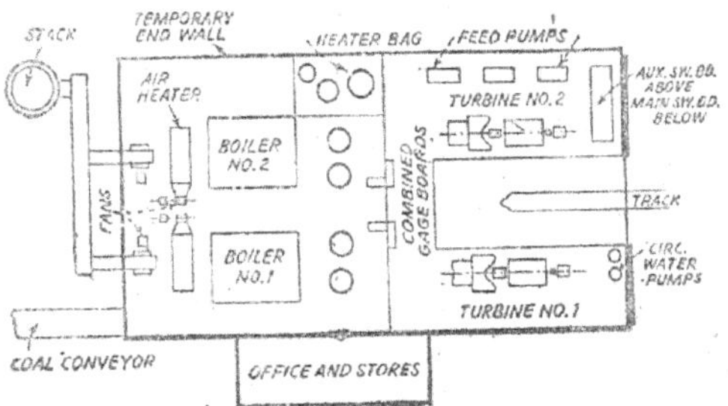

Fig. 2.2: Single Floor– station

In this arrangement, the boiler-operating floor is level with the furnace floor. The draught fans (ID and FD) and fed pumps are also placed on this floor. The turbine is located above on an island, but all controls are brought down to the main floor level. The switch gears for the generator and outgoing circuits are also placed on the same floor. All the piping is placed below the turbine gallery and practically hidden from view.

2.11.9. Unit Plan Station

With all modern developments in thermal sciences, materials and design, it has become customary to use u it plan even for high capacity power plants. This because, the single boiler and single turbine power plant is less expensive than multi boiler units. Presently unit boiler of 1000 to 2000 tons of steam per hour generating capacity and turbine generator sets of 100 to 500 MW capacities are available. Such arrangement permits the placement of turbines on the exact centreline of the boiler and facilities for the expansion of the power plant with this arrangement; one section does not extend beyond the other in the normal growth of the station. This also permits the duplication of units from identical design drawings. The arrangement of the plant is shown in fig 2.3.

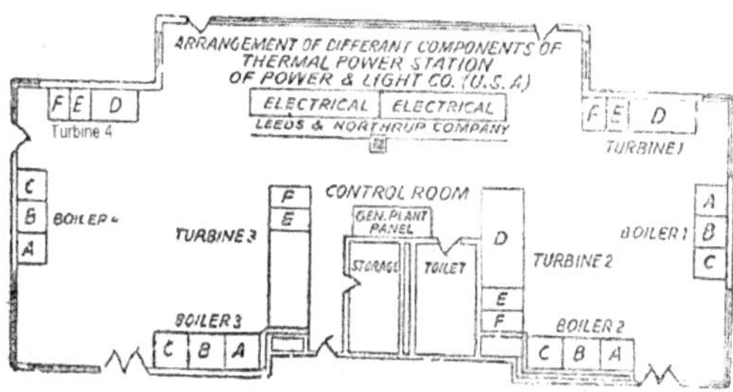

Fig. 2.3: Unit Plant Station

2.11.10.   Out –door type Steam Electric Station

Another type of arrangement which is commonly used in modern power plant layout is the out-door design. It is possible to save considerable cost of the plant buildings by an out-door layout. The out-door type of boiler arrangement was a natural development because the steel cased unit is equally satisfactory either indoors. The out-door boiler arrangement was first used in the south and the Gulf States of USA. The next development was to place certain portions of the electrical plant and heat cycle equipments of the turbine plant out of doors. Finally, the turbine generator unit has been placed out of doors either on deck or on foundation under a gantry crane. A fair weather-tight casing, even when located indoors normally encloses boiler, turbine generator sets, tanks, fans and other bulky equipments. Therefore, the modifications necessary to weatherproof such equipment are not expensive. The physical size of the building is considered reduced when out-door layout is used. It is established fact that 50 to 100 rupees per KW capacity, reduce the cost of plant by adopting outdoor arrangement. The layout of the system is shown in fig 2.4.

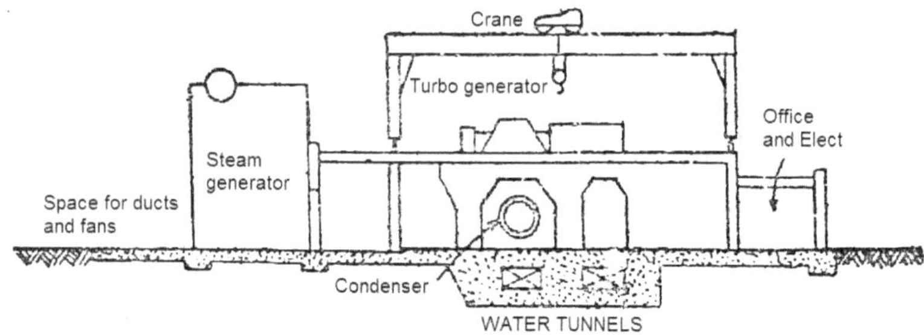

Fig. 2.4: Building Arrangement for An Outdoor Type Station

2.11.11.   Lighting of Steam Power Plants

The problem of illumination in a power station is somewhat specialized and the installations are totally different from the industrial lighting system for the reasons listed below

a. It is not necessary for the operator to examine the product closely.

b. Intensive local illumination is essential for instrument panels, gauge glasses and control boards.

c. High ceilings eliminate consideration for indirect lighting.

d. Fumes and moisture may corrode the lighting units.

e. Separate emergency lighting is absolutely essential because the power plant is the source of lighting service.

f. The coal handling and draught equipment make it difficult to illuminate the boiler room from above.

A power station operates 24 hours a day and 365 days a year and daylight is available only for about one-third of the running period. The sunlight is also not available in the night time; therefore, a provision for good artificial lighting should be made. The lighting provided should be adequate stead, evenly distributed for ensuring uniform lighting. In addition to this, the lighting system must ensure reliability of operation, safety, economy and ease of maintenance.

3. FUELS AND COMBUSTION

3.1. Introduction

India occupies 2.42% of the World's area and supports 16% of the World's population. It has embarked upon an ambitious program to build up its economy through Five Year Plans. In order to have a sustained economic growth, eight Five Year Plans have been implemented since India gained Independence. The Ninth Five Year Plan (1997-2002) envisages a growth rate of 7%. To achieve this rate of growth, the energy requirement for the country, would need to rise by 7 to 8 %.

Electricity is one of the most vital infrastructural inputs for the growth of the Indian economy. Since independence the Indian electricity sector has grown manifolds in size and capacity. The generating capacity has increased from a meager 1362 MW in 1947 to about 83,288 MW today. Electricity generation, which was only about 4.1 billion Kwh in 1947, has risen to a level of over 380 billion Kwh in 1995-96. The total length of transmission lines, today, is of the order of over 4 million ckt.kms. As compared to 29,271 ckt.kms in 1950. The numbers of villages which have been brought on the power map are over half a million, representing about 85% of the total villages in the country, as compared to only 3061 in 1950. Despite these impressive achievements, most regions in the country face severe power shortages. The per capita electricity consumption in the country, in spite of a rise to a level of 314 Kwh from 15.6 Kwh in 1950, is one of the lowest in the world.

In India, coal is the most abundant available fossil fuel and provides a substantial part of energy needs. It is used for power generation, to supply energy to industry as well as for domestic needs. India is highly dependent on coal for meeting its commercial energy requirements. Presently, about 62% of power generated in the country is from coal fired boilers. About 66% of the coal produced is consumed by Thermal Power Plants. This pattern is likely to continue. The coal based electricity generation capacity sharply increased from 8000 MW in 1970 to 51000 MW in 1995. This is expected to go up to 140000 MW by 2009-10. However, at present, the country faces an energy shortage of about 15% and peaking shortage of 30%.

Coal Mining in India was started in the year 1774 in the eastern part of the country in the State of West Bengal. At the beginning of this century, the total production of coal was just about 6 million tons per year. When India gained its independence in 1947, the coal production was nearly 30 million tons per year and the coal mining operation was primarily in the private sector. Till 1971-73 the coal mining operation remained primarily in the private sector and the production had come up to a level of nearly 72 million tons per year only. The entire coal industry in India was nationalized during 1972-73 and then on massive investments was made by the Government of India in this basic infrastructure sector. The production during 1996-97 was 286 million tons and coal production in 1997-98 would be nearly 298 mt. With this production growth, India now ranks as the third largest coal producer of the world next only to China and USA.

3.2. Fuel-Coal

Coal is created as the result of a natural chemical process in which vegetation is transformed by time, pressure and temperature. The time involved is very long to accommodate the organic chemical reaction which proceeds at a slow rate. Pressure is also important because coal of high rank is generally found in regions that have been under high pressure. The temperature at which the reaction takes place need not be high, for time brings about a relatively great change even at the lower temperatures prevailing in the earth's crust.

There is no satisfactory definition of coal. It is a mixture of organic chemistry and mineral materials-the organic chemical materials produce heat when burned and the mineral matter remains. An analysis, known as the **'proximate analysis'**, is used to rank coal and it determines four constituents in the coal:

1. Water, called moisture;
2. Mineral impurity, called ash, left when the coal is completely burned
3. Volatile matter, consisting of gases driven out when coal is heated to certain temperatures
4. Fixed carbon, the coke-like residue that burns at higher temperatures after the volatile matter has been driven off.

Coals are grouped according to rank and are known as anthracite, bituminous, sub bituminous and lignite. The coal rank increases as the amount of fixed carbon in the coal increases and ranges from anthracite (highest rank) to lignite (lowest rank).

3.2.1. Anthracite

Hard and very brittle, anthracite are dense and shiny black. It has a high percentage of fixed carbon and a low percentage of volatile matter. The amount of volatile matter in the coal influences the ease with which the coal can be burnt, with coals with a high amount of volatile matter being the easiest. This means that anthracite is difficult to burn and special consideration must be made in the design of the combustion system to ensure that stable combustion is achieved.

3.2.2. Bituminous

By far the largest group, bituminous coals derive their name from the fact that on being heated they are often reduced to a cohesive, binding, sticky mass. Their carbon content is less than that of anthracite, but they have more volatile matter and burn easily.

3.2.3. Sub Bituminous

These coals have high moisture content, as much as 15 to 30 percent, and are free-burning.

3.2.4. Lignite

Lignite is brown and of a laminar structure in which the remnants of wood fibers may be quite apparent. They have high volatile content and are free burning, but they have high moisture content (up to 65%) and low heating value so they are not economical to transport long distances. It is very important that the properties of the coal and ash are well known when designing a coal fired power station since they can strongly influence the capital cost and availability of the plant. In particular, coal properties influence the coal handling and boiler plants while ash properties influence the boiler, ash and dust handling and flue gas cleaning plants.

Extensive testing of the coal and ash is required prior to the design of a plant for a new coal field. This consists of laboratory and perhaps small scale tests, but usually not full scale due to the expense. The design of a plant for a new coal field is a difficult exercise since much of the design is reliant on empirical data from past experience on other similar coals. Sometimes this results in an inappropriate design due to an unusual coal/ash property which is not discovered until the plant is operational. Plant modifications (if feasible) are then required, which usually require an extensive plant outage resulting in lost revenue as well as capital expenses.

3.3. Use of Coal in Electricity Generation

The major use of coal in electricity generation is as a fuel burnt in the furnace of a large steam generator. The high pressure and temperature steam is then supplied by a turbo-generator which produces the electricity. The overall process is simple, but there is a large amount of associated plant and equipment used to optimize the cycle efficiency and minimize environmental pollution. The high usage of coal for electricity generation reflects the ready availability of low cost coal, which has enabled Australia to be one of the lowest cost producers of electricity in the world.

A more thermally efficient way to use coal is in an Integrated Gasification Combined Cycle (IGCC) plant. This is new technology and is not yet commercial, with only a few demonstration plants in the world. In an IGCC plant, coal is gasified and the gas burnt in combined cycle gas turbines. The coal may be gasified in a chemical reactor vessel that is integrated with the gas turbine plant. However, recent pulverized coal plants have been designed to operate at very high steam pressure and temperature with a much improved efficiency approaching that of IGCC without the complexity. Another method of coal gasification is to partially combust it while the coal is still underground. This is known as Underground Coal Gasification (UCG) and is an idea which has been around for over 100 years, but has not been adopted commercially in any Western economy. It has been used for about 40 years in Uzbekistan, where the gas was burnt in a conventional steam generator. The idea has been researched and trailed in many countries with the latest trials being conducted in Australia by CS Energy and Linc Energy. The proposal is to produce a low cost gas, which is burnt in a combined cycle gas turbine. The idea has some inherent practical difficulties concerning the monitoring of a partial combustion process underground together with the possibility of pollution of underground aquifers by the products of combustion.

3.3.1. *Inventory of Coal Reserves of India*

As a result of exploration carried out down to a depth of 1200m by the GSI and other agencies, a cumulative total of 245.69 Billion tonnes of coal resources has been estimated in the country as on 1.1.2004. The state-wise distribution of coal resources and its categorisation are shown in the table 3.1.

Table 3.1: State Wise Resources of Indian Coal

State	Coal Resources in Million Tonnes			
	Proved	Indicated	Inferred	Total
Andhra Pradesh	8091	6092	2514	16697
Arunachal Pradesh	31	40	19	90
Assam	279	27	34	340
Bihar	0	0	160	160
Chhattisgarh	8771	26419	4355	39545
Jharkhand	35409	30107	6348	71864
Madhya Pradesh	7513	8233	2914	18660
Maharashtra	4653	2156	1605	8414
Meghalaya	117	41	301	459
Nagaland	4	1	15	20
Orissa	14614	31239	15135	60988
Uttar Pradesh	766	296	0	1062
West Bengal	11383	11523	4488	27394
Total	**91631**	**116174**	**37888**	**245693**

3.3.2. *Categorization of Resources*

The coal resources of India are available in sedimentary rocks of older Gondwana formations of peninsular India and younger Tertiary formations of Northern/North-Eastern hilly region. Based on the results of Regional/Promotional Exploration, where the boreholes are placed 1-2 Km. apart, the resources are classified into Indicated or Inferred category. Subsequent Detailed Exploration in selected blocks, where boreholes are less than 400meter apart, upgrades the resources into more reliable Proved category. The Formation-wise and Category-wise coal resources of India as on 1.1.2004 are given in the table 3.2. and 3.3.

Table 3.2: The Formation-wise and Category-wise Coal Resources of India as on 1.1.2004

(in Million Tonnes)

Formation	Proved	Indicated	Inferred	Total
Gondwana Coals	91199	116068	37519	244786
Tertiary Coals	432	106	369	907
Total	**91631**	**116174**	**37888**	**245693**

Table 3.3: Type-wise and Category-wise Coal Resources of India as on 1.1.2004

(In Million Tonnes)

Type of Coal	Proved	Indicated	Inferred	Total
(A) Coking :				
-Prime Coking	4614	699	-	5313
-Medium Coking	11391	11774	1889	25054
-Semi-Coking	482	1003	222	1707
Sub-Total Coking	**16487**	**13476**	**2111**	**32074**
(B) Non-Coking*:	**75144**	**102698**	**35777**	**213619**
Total (Coking & Non-Coking)	**91631**	**116174**	**37888**	**245693**

Including coals of North Eastern Region.

3.3.3. *Oil*

Oil accounts for about 30% of India's total energy consumption. The majority of India's roughly 5.4 billion barrels in oil reserves are located in the Mumbai High, Upper Assam, Cambay, Krishna-Godavari, and Cauvery basins. The offshore Mumbai High field is by far India's largest producing field, with current output of around 260,000 barrels per day (bbl/d). India's average oil production level (total liquids) for 2003 was 819,000 bbl/d, of which 660,000 bbl/d was crude oil. India had net oil imports of over 1.4 million bbl/d in 2003.

Future oil consumption in India is expected to grow rapidly, to 2.8 million bbl/d by 2010, from 2.2 million bbl/d in 2003. India is attempting to limit its dependence on oil imports somewhat by expanding domestic exploration and production. To this end, the Indian government is pursuing the New Exploration Licensing Policy (NELP), first announced in 1997, which permits foreign involvement in exploration, an activity longer restricted to Indian state-owned firms. While the initial response to the 1999 tender was disappointing, with no bids received from the major multinational oil companies (causing an extension of the deadline for submission of bids), India proceeded with the award of 25 oil exploration blocks in early January 2000. The largest winner in the bidding round was India's domestic Reliance Industries, in partnership with independent Niko Resources of Canada, which received 12 blocks. British independent Cairn Energy, Russia's Gazprom, the U.S. firm Mosbacher Energy, and Geopetrol of France were all awarded single blocks in partnership with Indian firms. India's state-owned Oil and Natural Gas Corporation (ONGC) was awarded eight blocks, three of which it will hold in partnership with other public-sector Indian firms. A second round of bidding, with a total of 25 blocks offered, concluded in March 2001. Sixteen of the blocks have been awarded to ONGC, and four blocks to Hardy Oil of the United Kingdom, in partnership with India's Reliance Petroleum. The others were either awarded to smaller independent firms or failed to receive bids. As with the first round, no bids were received from major international oil companies. Bids for the third round were received in August 2002, with a total of 27 blocks offered. Awards under this third round were made in February 2003, with domestic Indian firms receiving most of the blocks. Reliance Industries received nine offshore blocks, one adjacent to the Krishna-Godavari Basin. ONGC was awarded 13 blocks, five offshore and eight onshore. The Gujarat State Petroleum Corporation received one. Blocks offered during the fourth round in 2003 received relatively little foreign interest. Awards for 15 blocks were made in February 2004, with 14 going to ONGC and one going to Reliance Industries. A sixth round of bidding opened in August 2004.

Low drilling recovery rates are a major part of the oil supply problem for India. Historically, recovery rates have averaged only around 30% in currently producing Indian oilfields, well below the world average. It is hoped that allowing foreign investment will bring in technology that is not available to Indian state firms, thereby increasing overall recovery rates. ONGC currently is undertaking a project to increase recovery rates in the Bombay High offshore field and several others as well, aiming to boost the overall recovery rate for its production assets from 28% to 40%. One area which has shown promise is western Rajasthan. Cairn Energy (UK) has been drilling in the area since 2001, and has reported several successful wells in 2004. The

Mangala field has been estimated to contain as much as 320 million barrels of recoverable reserves, and the "N-A" field has estimated recoverable reserves of 80 million barrels. Cairn is continuing exploration in the area, and is planning to bring the field into production by 2007, with an expected volume of 60,000 to 100,000 bbl/d.

In February 2002, BG purchased a 30% stake in the Panna, Mukta, and Tapti offshore oil and gas fields, which had previously been held by Enron. A dispute between BG and ONGC (which owns a 40% interest in the fields) over which firm would operate them was resolved in February 2003 with a "joint operatorship agreement." Reliance Industries holds the other 30% stake.

3.3.4. *Downstream/Refining*

For most of the 1990s, India imported a large quantity of refined products, lacking the refining capacity to keep up with growing demand. In 1999, refinery construction allowed India to close the gap. At the end of 2003, India had a total of 2.1 million bbl/d in refining capacity, an increase of 970,000 bbl/d since 1998. The largest single addition was Reliance Petroleum's huge Jamnagar refinery, which began operation in 1999. It has since reached its full capacity of 540,000 bbl/d. Jamnagar sells its products through three of the state-owned firms, and is in the process of building a retail network of its own, which is expected to include 2,000 retail outlets by the end of 2005.

Another major downstream infrastructure development is the construction of pipelines being undertaken by Petronet India, a company created by an agreement in 1998 between India's state-owned refineries. This construction is expected to add 500,000 bbl/d to India's current 325,000 bbl/d capacity for pipeline transportation of refined products. Pipelines between refineries and major urban centers are replacing rail cars as the main mode of transportation in India.

While state firms still control retail gasoline sales, several multinationals have entered the Indian lubricants market, which was deregulated five years ago. Firms such as Shell, ExxonMobil, and Caltex currently hold over one-third of the market. While these operations are relatively small, they are seen as allowing the majors to study the Indian market, establish brand recognition, and prepare for the eventual deregulation of the Indian retail petroleum products sector. Still, a requirement that foreign firms invest at least $400 million before entering the downstream market has served to limit their entry into petroleum products retailing. Shell met this requirement in early 2004, and intends to open a few retail outlets beginning in 2005.

3.3.5. Industry Restructuring and Price Deregulation

The Indian government officially ended the Administered Pricing Mechanism (APM) for petroleum product prices in April 2002. Prior to this deregulation, the Indian government had tried to offset the effects of price changes in crude oil by maintaining an Oil Pool Account, which was to build financial reserves when crude oil prices fell and release them back as increased subsidies when crude oil prices rose. In practice, though, the April 2002 reforms have not completely removed government influence on petroleum product prices. Subsidies have been maintained on some products, such as kerosene, which is commonly used as a cooking fuel by low-income households in India. State-owned downstream companies also still must submit proposed price changes to the Ministry of Petroleum and Natural Gas for approval. This has, in practice, limited movements in retail prices in response to fluctuations in world oil prices.

The previously planned sell off of government stakes in Hindustan Petroleum (HPCL) and Bharat Petroleum (BPCL) appear unlikely to move forward in the near future. The policy of the new Congress-led government is to avoid further privatizations of public companies which are making a profit. The new Congress-led government has reportedly been considering a restructuring of state-owned assets in the petroleum sector, which would consolidate IOC, ONGC, HPCL, and BPCL into two vertically-integrated major oil companies. No final decision has yet been made on such a restructuring.

India is planning to set up a strategic petroleum reserve equal to 15 days of the country's oil consumption. The state-owned refiner, Indian Oil Corporation (IOC) is likely to take the lead in the development of the reserve, which would be paid for by the Indian central government by means of a tax on petroleum product sales.

3.4. Natural Gas

Indian consumption of natural gas has risen faster than any other fuel in recent years. From only 0.6 trillion cubic feet (Tcf) per year in 1995, natural gas use was nearly 0.9 Tcf in 2002 and is projected to reach 1.2 Tcf in 2010 and 1.6 Tcf in 2015. A major development in December 2002 was the announcement by Reliance Industries of its discovery of a large amount of natural gas in the Krishna-Godavari Basin offshore from Andhra Pradesh along India's southeast coast. New reserves from this fund are estimated at about 7 Tcf. Reliance reported another find offshore from Orissa in June 2004, with estimated reserves of 1 Tcf. Cairn Energy also reported natural gas finds in late 2002 offshore from Andhra Pradesh as well as Gujarat, which contain reserves estimated at nearly 2 Tcf. The main market impacts from the new funds will be on India's east coast, which currently lacks extensive natural gas infrastructure.

Even with these new reserves, India's domestic natural gas supply is not likely to keep pace with demand, and the country will have to import much of its natural gas, either via pipeline or as liquefied natural gas (LNG). While the EIA's current forecast in the International Energy Outlook 2004 predicts a 4.8% annual growth rate in natural gas consumption, this reflects a substantial downward revision from previous forecasts, which had projected consumption of as much as 2.7 Tcf per year by 2010. Problems with financing LNG import projects have dimmed some of the previous prospects for explosive growth in natural gas consumption in India, and helped to revive interest in pipeline import options. Financial problems in the power sector, the main consumer of natural gas, also have had a negative effect.

Most of India's current natural gas production takes place in the Mumbai High basin and the state of Gujarat. Current projects include enhancing natural gas production at the Tapti fields in Gujarat and recovering previously flared natural gas at the Mumbai High oilfield.

India is investing heavily in the infrastructure required to support increased use of natural gas. Gas Authority of India Limited (GAIL), a government-owned entity, is in the process of doubling the throughput capacity on its main Hazira-Bijaipur-Jagdishpur (HBJ) Pipeline. Work on the capacity expansion began in 2002, and will eventually raise the capacity of the line from about 1.1 billion cubic feet per day (Bcf/d) to 2.1 Bcf/d. GAIL also plans a new distribution network in West Bengal and a pipeline which would connect Calcutta with Chennai. Shell has signed a memorandum of understanding with the state government of Uttar Pradesh in northern India for the development of a natural gas distribution infrastructure.

India's Foreign Investment Promotion Board (FIPB) had approved 12 prospective LNG import terminal projects in the mid-to-late-1990s, but it was never considered likely that all would be built in the near future, as their combined capacity would have exceeded even the most optimistic demand projections. The Indian government froze approvals of new LNG terminals in 2001, and payment problems at the Enron-backed Dabhol Power Plant in Maharashtra led many to question the financial viability of some of the LNG import projects. Reforms currently being undertaken in the electric power sector may eventually change this situation.

The largest state sector projects are to be conducted by Petronet, a joint venture between ONGC, IOC, the Gas Authority of India Ltd. (GAIL), the National Thermal Power Corporation (NTPC), and Gaz de France. Each of the state firms owns a 12.5% stake, the Gujarat state government owns a 5% stake, and the rest is owned by private investors, including a 10% stake held by Gaz de France Petronet plans two import terminals, one at Dahej and the other at Cochin. The import terminal at Dahej began operation earlier this year, receiving India's first cargo of LNG on January 30, 2004. The Dahej terminal had major advantages over some of the other proposed projects, because it is tied in with the main state-owned natural gas company, GAIL, and the existing HBJ pipeline network. Petronet is scheduled to start construction on its second terminal, at Kochi in Kerala state, in late 2005. The Shell also has begun construction of its LNG import terminal at Hazira in Gujarat, and has contracted for LNG supplies from Oman. The facility is scheduled to begin operation in November 2004. Like the Petronet Dahej terminal, it is to be linked into existing natural gas pipelines.

The Dabhol LNG terminal was nearly finished at the time construction was halted in June 2001, and it will likely be completed eventually, since construction was about 90% completed. Two American firms involved in the project, General Electric and Bechtel, purchased Enron's 65% stake in the project. At present, international arbitration is still pending over the financial terms of the project, mainly involving the government guarantees, and it is unclear when work on completing the facility will begin.

In the wake of the problems with Dabhol, firms backing several other LNG projects pulled out of India in the second half of 2001. Dhaksin Bharat Energy, a consortium including CMS Energy and Unocal, also announced the cancellation of its planned LNG project at Ennore. The Total has suspended further action on its planned LNG import terminal at Trombay. These LNG projects were cancelled largely in response to the Indian government's decision not to extend sovereign payment guarantees to power projects which were to have been among the import terminals' largest customers. Another proposed project in Andhra Pradesh on India's east coast may be jeopardized by cheaper natural gas supplies which will become available once Reliance Industries new offshore funds are developed. The BP-led consortium backing the project has switched the proposed location from Kakinada to Krishnapatnam, about 250 miles to the south.

Aside from LNG imports, imports of natural gas by pipeline may eventually play a role in satisfying India's gas needs. One possibility would supply India with natural gas from Iran's huge South Pars field via a pipeline, either subsea or through Pakistan. Iran has discussed the proposal with India and Pakistan. Australia's Broken Hill Proprietary (BHP) is the main foreign backer of the idea. An offshore route bypassing Pakistan also has been studied. Pakistan had said in early 2001 that it would allow supplies to cross its territory, and Iran would bear the contractual responsibility for assuring gas supplies to India. With the thaw in

India-Pakistan relations over the last year, the idea is again gaining some interest. Supplies of LNG from Iran might also be an option in the future, and IOC has opened discussions with the National Iranian Oil Company (NIOC) on a possible LNG export deal.

Another possible import route would link the natural gas reserves of Bangladesh into the Indian gas grid. Current proven reserves of natural gas in Bangladesh are at least 14 Tcf, but the foreign firms involved in natural gas exploration in Bangladesh, which includes Unocal, believe that reserves are higher. Shell, which backs exports to India, has estimated Bangladeshi natural gas reserves at 38 Tcf, and a study by the U.S. Geological Survey put the country's probable reserves at 32 Tcf. Bangladesh has been reluctant to approve exports to India, however, until all questions about reserves and its domestic supply have been resolved. After years of delays, Unocal effectively shelved the project in March 2004.

Finally, a new natural gas finds in Burma also has attracted interest as a potential source of supply for India. Indian companies ONGC and GAIL own a total of 30% equity in the reserves, and Bangladeshi officials stated in June 2004 that they would be willing to consider a pipeline running across Bangladeshi territory from Burma to West Bengal in India, provided agreement could be reached on terms and transit fees.

India's government has been considering reforms in its natural gas pricing mechanism, which is currently set by the government. Deregulation has been delayed several times, and buyers of natural gas from private sources such as the LNG terminal at Dahej pay prices much higher than those purchasing from the state-owned suppliers. With the shortage of natural gas and willingness of some consumers to pay more, deregulation would likely lead to higher prices if implemented.

3.5. Properties of Coal

3.5.1. Coal Classification

Coal is classified into three major types, namely anthracite, bituminous, and lignite. However, there is no clear demarcation between them and coal is also further classified as semi-anthracite, semi-bituminous, and sub-bituminous. Anthracite is the oldest coal from a geological perspective. It is a hard coal composed mainly of carbon with little volatile content and practically no moisture. Lignite is the youngest coal from a geological perspective. It is a soft coal composed mainly of volatile matter and moisture content with low fixed carbon. Fixed carbon refers to carbon in its free state, not combined with other elements. Volatile matter refers to those combustible constituents of coal that vaporize when coal is heated. The common coals used in Indian industry are bituminous and sub-bituminous coal. The gradation of Indian coal based on its calorific value is as follows:

The common coals used in Indian industry are bituminous and sub-bituminous coal. The gradation of Indian coal based on its calorific value is given in the table 3.4.

Table 3.4: The Gradation of Indian Coal Based on its Calorific Value

Grade	Calorific value Range(in k cal/kg)
A	Exceeding 6200
B	5600-6200
C	4940-5600
D	4200-4940
E	3360-4200
F	2400-3360
G	1300-2400

Normally D,E and F coal grades are available to Indian Industry.

The quality of coal depends upon its rank and grade. The coal rank arranged in an ascending order of carbon contents is

Lignite --> sub-bituminous coal --> bituminous coal --> anthracite

Indian coal is of mostly sub-bituminous rank, followed by bituminous and lignite (brown coal). The ash content in Indian coal ranges from 35% to 50%.

Chemical composition of the coal is defined in terms of its proximate and ultimate (elemental) analysis. The parameters of proximate analysis are moisture, volatile matter, ash, and fixed carbon. Elemental or Ultimate analysis encompasses the quantitative determination of carbon, hydrogen, nitrogen, sulfur, and oxygen. The calorific value Q, of coal is the heat liberated by its **complete** combustion with oxygen. Q is a complex function of the elemental composition of the coal. Gross Calorific value Q is mostly determined by experimental measurements. A close estimate can be made with the Dulong formula

$$Q = (144.4\ \%[C]) + (610.2\ \%[H]) - (65.9\ \%[O]) + (0.39\ \%[O]^2)$$

Q is given in Kcal/kg or Btu/lb. The Values of the elements C,H, and O, are calculated on a dry ash-free coal basis. The chemical composition of coal has a strong influence on its combustibility. The properties of coal are broadly classified as

1. Physical properties
2. Chemical properties

3.5.2. Physical Properties

3.5.2.1. Heating Value

The heating value of coal varies from coal field to coal field. The typical GCVs for various coals are given in the table 3.5.

Table 3.5: GCV for Various Coals

Parameter	Lignite (Dry Basis)	Indian Coal	Indonesian Coal	South African Coal
GCV(kcal/kg)	4.500	4.000	5.500	6.000

GCV of lignite on 'as received basis' is 2500-3000

3.5.2.2. Analysis of Coal

There are two methods: ultimate analysis and proximate analysis. The ultimate analysis determines all coal component elements, solid or gaseous and the proximate analysis determine only the fixed carbon, volatile matter, moisture and ash percentages. The ultimate analysis is determined in a properly equipped laboratory by a skilled chemist, while proximate analysis can be determined with a simple apparatus. It may be noted that proximate has no connection with the word "approximate".

3.5.2.3. Measurement of Moisture

Determination of moisture is carried out by placing a sample of powdered raw coal of size 200-micron size in an uncovered crucible and it is placed in the oven kept at 108+2 ºC along with the lid. Then the sample is cooled to room temperature and weighed again. The loss in weight represents moisture.

3.5.2.4. Measurement of Volatile Matter

A Fresh sample of crushed coal is weighed, placed in a covered crucible, and heated in a furnace at 900 + 15 ºC. For the methodologies, including that for carbon and ash, refer to IS 1350 part I: 1984, part III, IV. The sample is cooled and weighed. Loss of weight represents moisture and volatile matter. The remainder is coke (fixed carbon and ash).

3.5.2.5. Measurement of Carbon and Ash

The cover from the crucible used in the last test is removed and the crucible is heated over the Bunsen burner until all the carbon is burned. The residue is weighed, which is the incombustible ash. The difference in weight from the previous weighing is the fixed carbon. In actual practice Fixed Carbon or FC derived by subtracting from 100 the value of moisture, volatile matter and ash.

3.5.2.6. Proximate Analysis

Proximate analysis indicates the percentage by weight of the Fixed Carbon, Volatiles, Ash, and Moisture Content in coal. The amounts of fixed carbon and volatile combustible matter directly contribute to the heating value of coal. Fixed carbon acts as a main heat generator during burning. High volatile matter, content indicates easy ignition of fuel. The ash content is important in the design of the furnace grate, combustion volume, pollution control equipment and ash handling systems of a furnace. A typical proximate analysis of various coal is given in the Table 3.6. Determines (on an as-received basis)

- **Moisture content**
- **Volatile matter** (gases released when coal is heated).
- **Fixed carbon** (solid fuel left after the volatile matter is driven off).
- **Ash** (impurities consisting of silica, iron, alumina, and other incombustible matter).

Table 3.6: Typical Proximate Analysis of Various Coals (in percentage)

Parameter	Indian Coal	Indonesian Coal	South African Coal
Moisture	5.98	9.43	8.5
Ash	38.63	13.99	17
Volatile matter	20.70	29.79	23.28
Fixed carbon	34.69	46.79	51.22

3.5.3. Chemical Properties

3.5.3.1. Ultimate Analysis

The ultimate analysis indicates the various elemental chemical constituents such as Carbon, Hydrogen, Oxygen, Sulphur, etc. It is useful in determining the quantity of air required for combustion and the volume and composition of the combustion gases. This information is required for the calculation of flame temperature and the flue duct design, etc. Typical ultimate analyses of various coals are given in the Table 3.7.

Table 3.7: Typical Ultimate Analysis of Coals

Parameter	Indian Coal in %	Indonesian Coal IN %
Moisture	5.98	9.43
Mineral Matter(1.1xAsh)	38.63	13.99
Carbon	41.11	58.96
Hydrogen	2.76	4.16
Nitrogen	1.22	1.02
Sulphur	0.41	0.56
Oxygen	9.89	11.88

Indian coal is classified by grades[8] defined on the basis of Useful Heat Value (UHV). UHV is an expression derived from ash and moisture contents for non-cocking coals as per the Government of India notification. The ultimate analysis of coal used in power plants in India is readily not available. Ultimate analysis of D, E, and F grade coal is obtained as personal communication from Central Fuel Research Laboratory (CFRI), Jharkhand, India. Ultimate analysis of coal used at Dadri, Rihand, Singrauli, Chandrapur, and Dahanu power plants is obtained as personal communication from the National Energy Technology Laboratory, Pittsburgh, USA. The Neyveli Lignite Corporation provided by personal communication, the ultimate analysis for lignite used at the Neyveli and Kutch power plants. These are presented in Tables 3.8A & 3.8B.

Table 3.8 A: Elemental Analysis, Moisture Content, and Grades of Typical Indian Coals

Coal Grade	C%	H%	S%	N$_2$%	O$_2$%	A%	M%	NCV (Kcal/Kg)	UHV (cal/gm)
D	33.1	2.46	0.44	0.83	NA	25.9	7.2	4999.0	4332.0
D	30	2.48	0.57	0.69	NA	27.1	2.9	5555.0	4760.0
D	32.31	2.12	0.4	0.78	NA	25	7.3	5068.0	4442.0
E	37.9	2.4	0.53	0.8	6%	30.4	7.5	4529.0	3670.0
F1	41.87	3.33	0.56	0.94	6%	34.07	7.8	4137.0	3122.0
F2	44.47	3.37	0.35	0.99	6%	36.3	8.4	3833.0	2731.0
Average of E and F2	41.19	2.89	0.44	0.9	0.06	33.35	7.95	4182.0	-

Table 3.8 B: Elemental Analysis and Moisture Content of the Coal Used at the Seven Power Plants in India

Coal	C%	H%	S%	N$_2$%	O$_2$%	A%	M%	NCV
Dadri	40.3	4.16	0.5	0.9	15.92	38.22	NA	NA
Rihand	37.74	3.26	0.39	0.73	14.65	43.23	NA	NA
Singrauli	50.22	4.78	0.33	1.09	17.25	26.33	NA	NA
Chandrapur	37.69	2.66	0.8	1.07	5.78	47.0	5	3649.9
Dahanu	42.39.0	3.73	0.39	0.82	14.21	38.46	5.93	3986.37
Nevyeli Lignite	26.09	2.33	1.5	0.24	16.33	7.0	47	2229.0
Kutch Lignite	28.33	3.03	2.25	0.88	13.94	15.0	36	2900.0

NA: Not Available

C: Carbon

H: Hydrogen

S: Sulfur

N: Nitrogen

CV: Calorific value

A: Ash

M: Moisture

UHV: Useful heat value = 8900 - 138(A+M)

GCV: Gross Calorific Value = (UHV + 3645 -75.4 M)/1.466

NCV: Net Calorific Value = GCV - 10.02M

The Relationship of GCV, UHV, and NCV is empirical.

Ultimate analysis of typical United States coals[9] is given in Table 3.9 for the sake of comparison.

Table 3.9: Ultimate Analysis of Typical US Coals

Coal	C%	H%	S%	N$_2$%	O$_2$%	A%	M%	NCV (Kcal/Kg)
Pennsylvania	65.8	4.6	2.3	1.4		19.8	1.1	6567
Ohio	64.2	5.0	1.8	1.3		16.0	2.8	6378
West Virginia	72.1	4.8	1.0	1.4		11.7	1.8	7025
Kentucky	70.9	5.1	2.3	1.5		9.8	2.3	7022
West Virginia	70.0	5.1	1.2	1.5		8.1	1.2	7522
Illinois	77.4	5.5	2.5	1.4		8.2	4.1	7027
Pennsylvania	77.4	5.2	2.4	1.4		7.5	1.1	7728
Illinois	73.7	5.1	2.3	1.6		7.9	2.0	7330

The Importance of the quality of coal is illustrated by the comparison of coal used at Chandrapur Thermal Power Plant (India) and the Ohio (USA) coal. The ultimate analyses of the two coal types are given in Table 3.10. The Calorific value of Ohio coal is almost twice to that of Chandrapur coal. This means that roughly twice the Indian coal compared to Ohio coal is needed to generate the same quantity of steam (electricity). Assuming 30% thermal conversion efficiency in converting thermal energy from coal to the electrical energy, coal used for generating a unit of electricity is (Kg/KWH) 0.77 for Indian coal and 0.36 for Ohio coal.

Table 3.10: Comparison (Ultimate analysis) of Chandrapur and Ohio coal

		Chandrapur (India)	Ohio (USA)
1	Fixed carbon	27.5%	44.0
2	Total carbon	37.69%	64.2%
3	Hydrogen	2.66%	5.0 %
4	Nitrogen	1.07%	1.3 %
5	Oxygen (difference)	5.78%	11.8 %
6	Sulfur	0.8%	1.8 %
7	Ash	47%	16%
8	Total moisture	5%	2.8
9	Gross calorific value Kcal/Kg	3400	6378
10	Coal per unit of electricity (Kg/KWH)	0.77	0.36

The coal properties including calorific values differ depending upon the colliery. The calorific value of the Indian coal (~15 MJ/kg) is less than the normal range of 21 to 33 MJ/Kg (gross). The design rating of a coal-fired burner (in USA) is at 26 MJ/Kg.

3.6. Combustion

3.6.1. *Principle of Combustion*

Combustion refers to the rapid oxidation of fuel accompanied by the production of heat, or heat and light. Complete combustion of a fuel is possible only in the presence of an adequate supply of oxygen. Oxygen (O2) is one of the most common elements on earth,, making up 20.9% of our air. Rapid fuel oxidation results in large amounts of heat. Solid or liquid fuels must be changed to a gas before they will burn. Usually heat is required to change liquids or solids into gases. Fuel gases will burn in their normal state if enough air is present. Most of the 79% of air (that is not oxygen) is nitrogen, with traces of other elements. Nitrogen is considered to be a temperature, reducing dilution that must be present to obtain the oxygen required for combustion. Nitrogen reduces combustion efficiency by absorbing heat from the combustion of fuels and diluting the flue gases.

This reduces the heat available for transfer through the heat exchange surfaces. It also increases the volume of combustion by-products, which then have to travel through the heat exchanger and up the stack faster to allow the introduction of additional fuel air mixture. This nitrogen also can combine with oxygen (particularly at high flame temperatures) to produce oxides of nitrogen (NOx), which are toxic pollutants. Carbon, hydrogen and sulphur in the fuel combine with oxygen in the air to form carbon dioxide, water vapour and sulphur dioxide, releasing 8084 kcals, 28922 kcals & 2224 kcals of heat respectively. Under certain conditions, Carbon may also combine with Oxygen to form Carbon Monoxide, which results in the release of a smaller quantity of heat (2430 kcals/kg of carbon) Carbon burned to CO_2 will produce more heat per pound of fuel than when CO or smoke are produced.

$C + O_2 \rightarrow CO_2 + 8084$ kCals/kg of Carbon

$2C + O_2 \rightarrow 2CO + 2430$ kCals/kg of Carbon

$2H_2 + O_2 \rightarrow 2H_2O + 28,922$ kCals/kg of Hydrogen

$S + O_2 \rightarrow SO_2 + 2,224$ kCals/kg of Sulphur

Each kilogram of CO formed means a loss of 5654 kCal of heat.(8084-2430).

3.6.2. 3 T's of Combustion

The objective of good combustion is to release all of the heat in the fuel. This is accomplished by controlling the "three T's" of combustion which are (1) Temperature high enough to ignite and maintain ignition of the fuel, (2) Turbulence or intimate mixing of the fuel and oxygen, and (3) Time sufficient for complete combustion.

Commonly used fuels like natural gas and propane generally consist of carbon and hydrogen. Water vapour is a by-product of burning hydrogen. This robs heat from the flue gases, which would otherwise be available for more heat transfer. Natural gas contains more hydrogen and less carbon per kg than fuel oils and as such produces more water vapour. Consequently, more heat will be carried away by exhaust while firing natural gas.

Too much, or too little fuel with the available combustion air may potentially result in unburned fuel and carbon monoxide generation. A very specific amount of O_2 is needed for perfect combustion and some additional (excess) air is required for ensuring complete combustion. However, too much excess air will result in heat and efficiency losses.

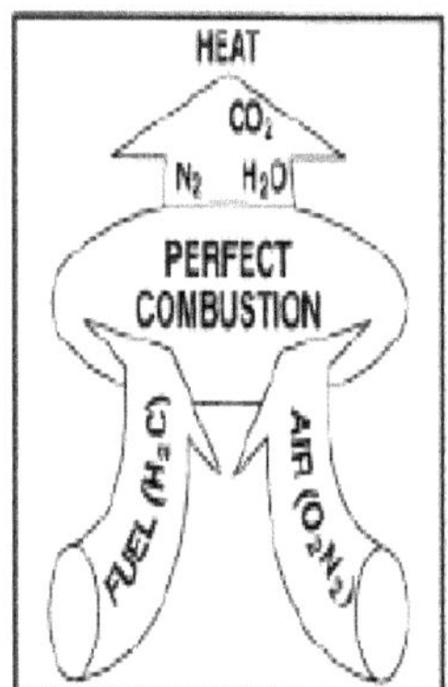

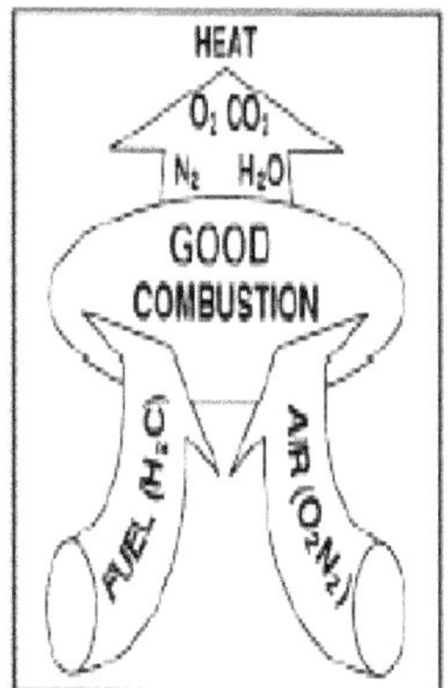

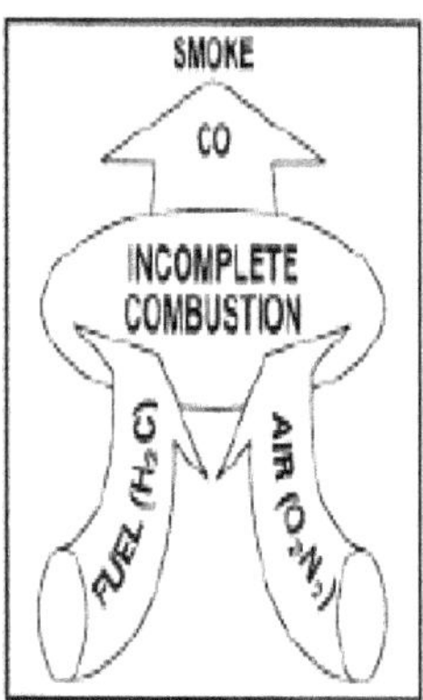

Fig. 3.1: 3 T's of Combustion

Not all of the heat in the fuel are converted to heat and absorbed by the steam generation equipment. Usually all of the hydrogen in the fuel is burned and most boiler fuels, allowable with today's air pollution standards, contain little or no sulfur. So the main challenge in combustion efficiency is directed toward unburned carbon (in the ash or incompletely burned gas), which forms CO instead of CO_2.

3.7. Emissions from Coal Usage

The main emissions from coal combustion at thermal power plants are Carbon dioxide (CO_2), Nitrogen oxides (NO_x), Sulfur oxides (SO_x), Chlorofluorocarbons (CFCs), carbonaceous material (soot), and air-borne inorganic particles such as fly ash, also known as suspended particulate matter (SPM) and other trace gas species. Carbon dioxide, nitrous oxide, and chlorofluorocarbons are greenhouse gases. Evidence accumulated by the Inter-governmental Panel on Climate Change (IPCC) suggests that emissions of these greenhouse gases might be responsible for climate change, a global concern. Possible consequences projected by IPCC include: a rise in sea levels-a more vigorous hydrological cycle that may increase the severity of floods and droughts and may cause more extreme climatic events; and - ecological change that could threaten agricultural productivity.

Oxides of nitrogen and sulfur, also play an important role in atmospheric chemistry and are largely responsible for atmospheric acidity. Particulates and black carbon (soot) are of concern in the radiative[10] forcing of the earth. They also have a significant negative impact on human health, causing lung tissue irritation and are linked to cancer and other serious diseases.

The pollutants emitted from thermal power plants depend largely upon the fuel burned, the furnace design, the excess air, and any additional devices used to reduce the emissions. At present, the only control device used in thermal power plants in India is an electrostatic precipitator to control the emission of fly ash (SPM). CO_2, SO_2, nitric oxide (NO), soot, and SPM emissions from each of the thermal (coal-fired) power plants in India have been computed using basic principles of combustion. These calculations are based on a theoretical ideal and the input data, such as chemical composition of the coal used in the power plants, coal used per unit of power, excess air used during combustion, and the power generation from each plant..

3.8. Emission of Carbon Dioxide and Sulfur Dioxide

Utilities mostly burn coal with approximately 10 -30% excess air. Carbon as obtained from Ultimate analysis is converted to CO_2 after the reaction (combustion) is complete. Some carbon is emitted in the form of soot and some carbon remains unburned and mixes with the ash. Different combustion technologies affect the types and concentrations of resultant species e.g. fixed bed combustion results in higher carbon content in the ash. Carbon in the coal is converted to carbon dioxide (CO_2) by the reaction

$$C + O_2 \rightarrow CO_2$$

Similarly, hydrogen and sulfur are converted to moisture (H_2O) and sulfur dioxide (SO_2) by the reactions

$$H_2 \quad + \quad O_2/2 \quad \rightarrow \quad H_2O$$
$$S \quad + \quad O_2 \quad \rightarrow \quad SO_2$$

Table 3.11, as an example, gives the computation of oxygen required for burning one kg of coal used at the Chandrapur thermal power station and the combustion products.

Table 3.11: Computation of Combustion Products for the Chandrapur Coal

Species	Mass	Oxygen required	Products	Reaction
Carbon	0.3769	0.3769 x (32/12) = 1.01	CO_2: 0.3769 x (44/12) = 1.38	$C + O_2 \dashrightarrow CO_2$
Hydrogen	0.0266	0.0266 x (16/2) = 0.21	H_2O: 0.0266 x (18/2) = 0.24	$H_2 + O_2/2 \dashrightarrow H_2O$
Sulphur	0.008	0.008 x (32/32) = 0.008	SO_2: 0.008 x (64/32) = 0.016	$S + O_2 \dashrightarrow SO_2$
Oxygen	0.0578			
Nitrogen	0.0107			$N_2 + O_2 \dashrightarrow 2NO$
Ash	0.47			

Oxygen required to burn 1 Kg of coal=1.01+0.21+0.01−0.05=1.18 Kg

Air required = (Oxygen required)/(Mass fraction of oxygen in the air) = 1.18/0.233 = 5.06 Kg = stoichiometric air

Total air (stoichiometric + 20% excess) = 5.06x1.2 = 6.072 Kg

6.072 air = 1.415(6.072 x 0.233) oxygen + 4.474(6.072 x 0.767) nitrogen

The combustion product with 20% excess air will contain:

0.2348(1.415 – 1.18)Kg O_2 and 4.48(4.47 + 0.01)Kg N_2.

Appendix-A gives the general formula for the calculation of required oxygen for burning of one kg of fuel.

Table 3.12 gives the computation of the flue gas composition from burning one Kg of Chandrapur coal at 20% excess air, as an example.

Table 3.12: Flu Gas Composition with 20% Excess Air

Product	Mass/Kg coal	Mol. wt	Kmoles/Kg coal	% volume
CO_2	1.38	44	0.03136	15.39
SO_2	0.016	64	0.00025	0.12
O_2	0.24	32	0.0075	3.68
N_2	4.61	28	0.1646	80.77
		Total =	0.2038	

3.9. Emissions of Oxides of Nitrogen from Coal

Oxides of nitrogen (NO_x) are (i) nitrous oxide (N_2O), (ii) nitric oxide (NO), and (iii) nitrogen dioxide (NO_2). NO_2 is mostly formed by oxidation of the NO, which is discharged in combustion products. About 90% of the NO_x is in the form of NO. NO is formed by two mechanisms: (i)

oxidation of atmospheric nitrogen, known as 'thermal NO' and (ii) oxidation of nitrogen that is chemically bound within the fuel, known as 'chemical NO'. The amount of NO_x varies widely with boiler conditions. NO_x emissions are generally functions of flame temperature, excess air, percentage of boiler load, nitrogen content in the coal, and rate of gas cooling. In pulverized coal flames, about 30-35% of nitrogen in coal gets converted into NO and the remaining nitrogen in the coal gets converted into molecular nitrogen. The actual mechanism, whereby atmospheric nitrogen is oxidized, goes through a complex chain of reactions initiated by oxygen atoms. We can, however calculate equilibrium concentrations of NO, using the following reaction:

$$N_2 \quad + \quad O_2 \quad \text{-->} 2NO$$

This is a lumped reaction. Generally accepted principal reactions are

$$O + N_2 = NO + N$$
$$N + O_2 = NO + O$$
$$N + OH = NO + H$$

The concentration of nitric oxide (NO) is given by

$$X_{NO} = K_{10.1} \left(X_{N_2}\right)^{0.5} \left(X_{O_2}\right)^{0.5}$$

Where X is the species concentration and $K_{10.1}$ is a equilibrium constant and depends upon the temperature of the gas. Appendix-B gives the equations to compute equilibrium constant for NO reaction. Concentration values X for O_2, N_2, CO_2, SO_2, and NO for the Chandrapur coal, as calculated by the above method, are given in Table 3.13. The value of the equilibrium constant $K_{10.1}$ computed at 1700 K is 0.007824.

Table 3.13: Species Concentrations in Flue Gas from the Sample Coal

Air	X(O_2)	X(N_2)	X(CO_2)	X(SO_2)	X(NO)
Stoichiometric	0	0.831333	0.167148	0.001512	0
5% excess	0.010483	0.8293267	0.158754	0.001436	0.00073
10% excess	0.019970	0.8274488	0.151213	0.001368	0.001006
15% excess	0.028597	0.8257412	0.144356	0.001306	0.001202
20% excess	0.036475	0.8241817	0.138094	0.001249	0.001357
25% excess	0.043699	0.8227519	0.132352	0.001197	0.001484
30% excess	0.050345	0.821436	0.127069	0.001149	0.001591

Figure 3.2 gives a comparative behavior of the CO_2, SO_2, and NO concentrations (in mole fraction) for 5-30% excess air. NO concentrations increase, while the concentrations of CO_2 and SO_2 decrease with the increase in excess air.

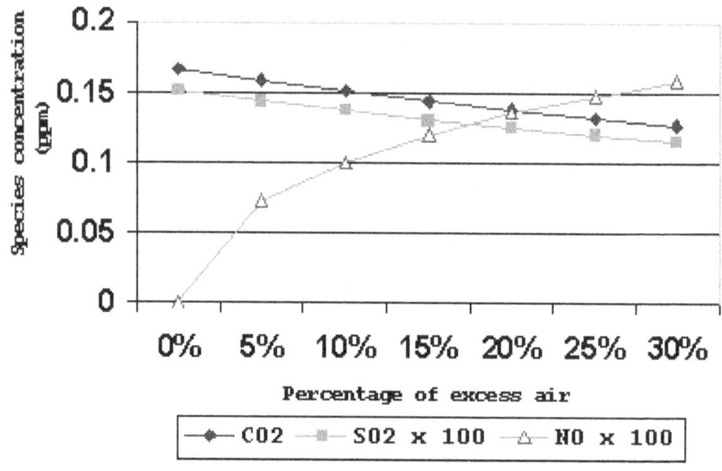

Fig. 3.2: Combustion product per kg of coal

Species concentrations (PPM) in the combustion products for the Chandrapur coal are given in Table 3.14.

Table 3.14: Species Concentrations in Parts Per Million (ppm)

Excess Air %	CO_2	SO_2	NO
0	167148	1512	0
5	158754	1436	730
10	151213	1368	1006
15	144356	1306	1202
20	138094	1249	1357
25	132352	1197	1484
30	127069	1149	1591

3.10. Carbonaceous Material and Black Carbon

Incomplete and/or inefficient combustion processes of fossil fuel generate carbonaceous aerosols. The emitted carbonaceous (soot) aerosols are of two types, namely organic carbon (OC) and black carbon (BC). These two types have different properties in the atmosphere. OC is a reactive species and has scattering properties in the solar spectrum[12] while the BC, on the other hand, is non reactive in the atmosphere but has highly absorbing properties in the solar spectrum. In the thermal power plants, most of the soot, carbon emitted would be in the form of BC because of the high temperatures of combustion in the furnaces. When the soot is formed the analysis ranges from C_8H to $C_{12}H$. The importance of BC in the radiative balance of the Earth is gradually being understood.

The soot, carbon from the combustion of coal in the Indian thermal power plants has been calculated on the basis of prevalent combustion characteristics. It is assumed that approximately, 10% of the coal carbon goes in the bottom ash and about 2% of the carbon forms the soot that may emit with fly ash. Electrostatic Precipitators (ESPs) used in the thermal power plants in India, remove about 99% fly ash from the stack. 1% of the generated fly ash, which contains 2% soot, carbon, is emitted into the atmosphere. These soot particles are sub-microns in size. Similar assumption has also been used for calculation of soot, carbon from lignite based thermal power plants in India (viz. Kutch and Neyveli power plants).

3.11. Suspended Particulate Matter

The fly ash in the form of Suspended Particulate Matter (SPM) is a major pollutant from coal burning power plants in India. SPM has been calculated on the basis of the ash contents of the coal. It is assumed that 85% of the ash in the coal goes out through the stack as fly ash after the combustion. Electrostatic precipitators (ESPs) working on 99% efficiency rate allows only 1% of the formed fly ash to emit as SPM.

3.12. Optimizing Excess Air and Combustion

For complete combustion of every one kg of fuel oil 14.1 kg of air is needed. In practice, mixing is never perfect, a certain amount of excess air is needed to complete combustion and ensure that release of the entire heat contained in fuel oil. If too much air than what is required for completing combustion were allowed to enter, additional heat would be lost in heating the surplus air to the chimney temperature. This would result in increased stack losses. Less air would lead to the incomplete combustion and smoke. Hence, there is an optimum excess air level for each type of fuel.

3.13. Control of Air and Analysis of Flue Gas

Thus, in actual practice, the amount of combustion air required will be much higher than optimally needed. Therefore, some of the air gets heated in the furnace, boiler and leaves through the stack without participating in the combustion Chemical analysis of the gases is an objective method that helps in achieving finer air control. By measuring carbon dioxide (CO_2) or oxygen (O_2) in flue gases by continuously recording instruments or Orsat apparatus or portable fyrite, the excess air level as well as stacks losses can be estimated with the graph as shown in Figure 3.3 and Figure 3.4 The excess air to be supplied depends on the type of fuel and the firing system. For optimum combustion of fuel oil, the CO_2 or O2 in flue gases should be maintained at 14 -15% in the case of CO_2 and 2-3% in the case of O_2.

3.14. Combustion of Coal

3.14.1. Coal-fired Burners

Now we are getting to the larger burners! Coal-fired burners are in common use in heavy industry, especially in countries with reserves of coal. They can present a problem for our flue gas analyzer due to the amount of smoke and other pollutants that some forms of coal can produce. There are very many different varieties and qualities of coal, ranging from peat to high quality material that produces a high level of energy and less waste. The efficiency of the combustion will depend largely on the preparation of the coal before burning. Power station burners will pulverize the coal and blow it in as a fine dust. The smallest burners will simply feed the chunks of coal as delivered into the combustion chamber. Obviously, the mixing of fuel and air is much better with pulverized fuel and the combustion will be both quicker and cleaner, as the flue gas analyzer will show.

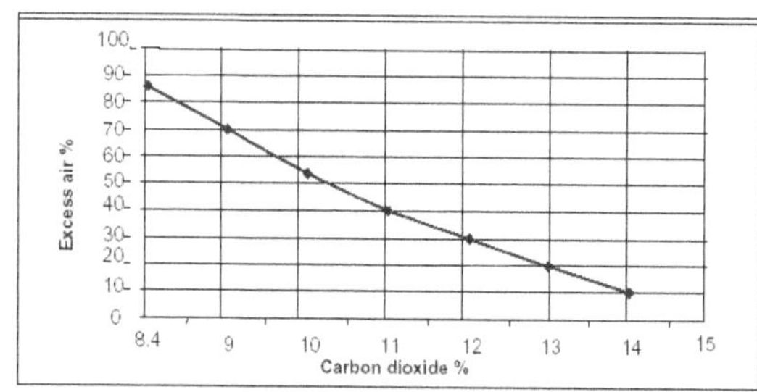

Fig. 3.3: Relations between CO_2 and Excess Air for Fuel

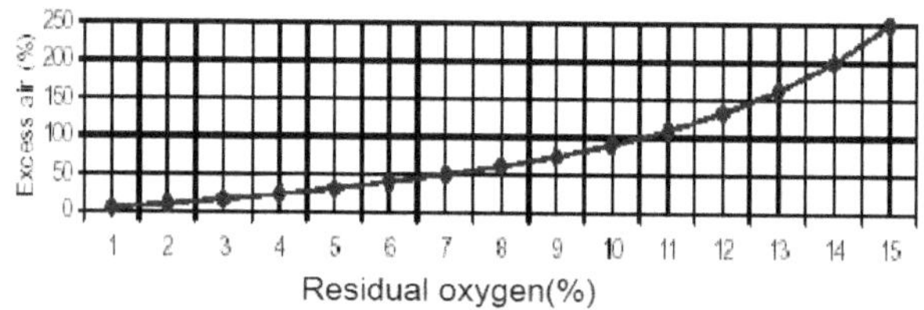

Fig. 3.4: Relation between Residual Oxygen and Excess Air

Coal may contain a large amount of sulphur and other impurities, which will be found in the flue gas when measured. Particularly inferior brands of coal will contain tarry matter that will cause sticky smoke that can block a filter on a flue gas analyzer very quickly. Care must be taken to inspect the filters and change them if necessary. A blocked filter may also lead to air being drawn in at the joints in the gas tubing, which falsifies the readings on the flue gas analyser. Coal burners require a flue gas analyzer capable of measuring sulphur dioxide. This will always be present to some degree. The level of nitrogen dioxide will also be elevated due to nitrogen bearing compounds in the coal. This means that a calculation of NOx from nitric oxide will not be reliable and a nitrogen dioxide sensor should really also be fitted. Water is less of a problem here, except with some forms of brown coal or peat, which may contain free water as well, boosting the level of condensate greatly. The main problems are smoke and tarry deposits. A soot test will be essential and extra filtration for the flue gas analyzer should be considered in many cases.

1 kg of coal will typically require 7-8 kg of air depending upon the carbon, hydrogen, nitrogen, oxygen and sulphur content for complete combustion. This air is also known as theoretical or stoichiometric air. If for any reason the air supplied is inadequate, the combustion will be incomplete. The result is poor generation of heat with some portions of carbon remaining unburnt (black smoke) and forming carbon monoxide instead of carbon dioxides.

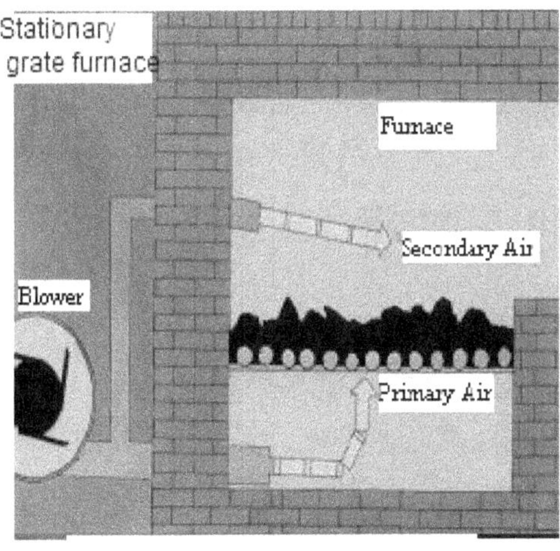

Fig. 3.5: Coal Combustion

As in the case of oil, coal cannot be burnt with the stoichiometric quantity of air. Complete Combustion is not achieved unless an excess of air is supplied. The excess air required for coal combustion depends on the type of coal firing equipment. Hand fired boilers use large lumps of coal and hence need very high excess air. Stoker fired boilers as shown in the Figure 3.5 use sized coal and hence requires less excess air. Also in these systems primary air is supplied below the grate and secondary air is supplied over the grate to ensure complete combustion. Fluidized bed combustion in which turbulence is created leads to intimate mixing of air and fuel, resulting in further reduction of excess air. The pulverized fuel firing in which powdered coal is fired has the minimum excess air due to high surface area of coal ensuring complete combustion.

4. STEAM GENERATORS

4.1. Introduction

A **boiler** is an enclosed vessel that provides a means for combustion heat to be transferred into water until it becomes heated water or steam. The hot water or steam under pressure is then usable for transferring the heat to a process. Water is a useful and cheap medium for transferring heat to a process. When water is boiled into steam its volume increases about 1,600 times, producing a force that is almost as explosive as gunpowder. This causes the boiler to be extremely dangerous equipment that must be treated with utmost care. The process of heating a liquid until it reaches its gaseous state is called evaporation. Heat is transferred from one body to another by means of (1) radiation, which is the transfer of heat from a hot body to a cold body without a conveying medium, (2) convection, the transfer of heat by a conveying medium, such as air or water and (3) conduction, transfer of heat by actual physical contact, molecule to molecule.

4.2. Boiler Specification

The heating surface is any part of the boiler metal that has hot gases of combustion on one side and water on the other. Any part of the boiler metal that actually contributes to making steam is heating surface. The amount of heating surface of a boiler is expressed in square meters. The larger the heating surface a boiler has, the more efficient it becomes. The quantity of the steam produced is indicated in tons of water evaporated to steam per hour. Maximum continuous rating is the hourly evaporation that can be maintained for 24 hours. F & A means the amount of steam generated from water at 100oC to saturated steam at 100 oC

Typical Boiler Specification

Boiler Make & Year	: XYZ & 2003
MCR (Maximum Continuous Rating)	:10TPH (F & A 100oC)
Rated Working Pressure	: 10.54 kg/cm^2(g)
Type of Boiler	: 3 Pass Fire tube
Fuel Fired	: Fuel Oil

4.3. Boiler Systems

The boiler system comprises of: feed water system, steam system and fuel system. The feed water system provides water to the boiler and regulates it automatically to meet the steam demand. Various valves provide access for maintenance and repair. The steam system collects and controls the steam produced in the boiler. Steam is directed through a piping system to the point of use. Throughout the system, steam pressure is regulated using valves and checked with steam pressure gauges. The fuel system includes all equipment used to provide fuel to generate the necessary heat. The equipment required in the fuel system depends on the type of fuel used in the system. A typical boiler room schematic is shown in Figure 4.1.

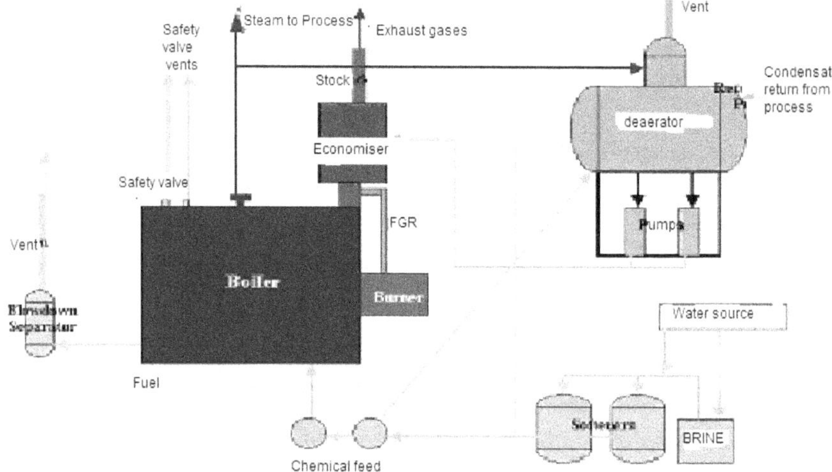

Fig. 4.1: Boiler Room

The water supplied to the boiler that is converted into steam is called feed water. The two sources of feed water are: (1) Condensate or condensed steam returned from the processes and (2) Makeup water (treated raw water) which must come from outside the boiler room and plant processes. For higher boiler efficiencies, the feed water is preheated by economizer, using the waste heat in the flue gas.

4.4. Boiler Types and Classifications

There are virtually infinite numbers of boiler designs but generally they fit into one of two categories:

4.4.1. Fire Tube

Fire tube or "fire in tube" boilers; contain long steel tubes through which the hot gasses from a furnace pass and around which then water to be converted to steam circulates. (Refer Figure 4.2). Fire tube boilers, typically have a lower initial cost, are more fuel efficient and easier to operate, but they are limited generally to capacities of 25 tons/hr and pressures of 17.5 kg/cm2.

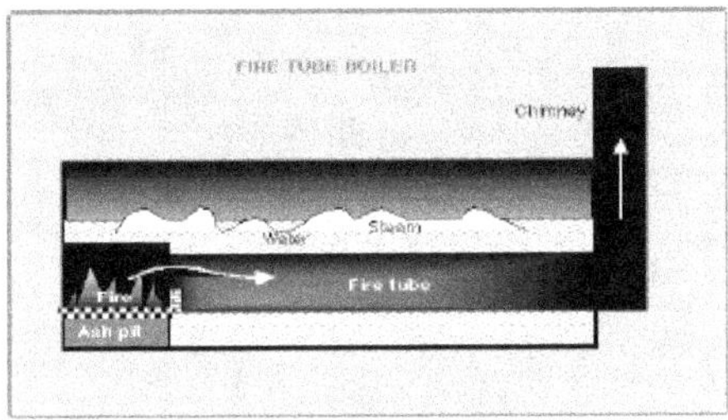

Fig. 4.2: Fire Tube Boiler

4.4.2. Water Tube

Water tube or "water in tube" boilers in which the conditions are reversed with the water passing through the tubes and the hot gasses passing outside the tubes (see figure 4.3). These boilers can be of single- or multiple-drum type. These boilers can be built to any steam capacities and pressures, and have higher efficiencies than fire tube boilers.

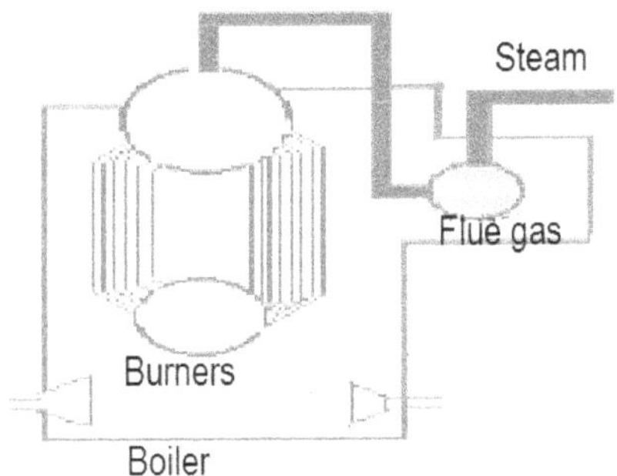

Fig. 4.3: Water Tube Boiler

4.4.3. Packaged Boiler

The packaged boiler is so called because it comes as a complete package. Once delivered to site, it requires only the steam, water pipe work, fuel supply and electrical connections to be made for it to become operational. Package boilers are generally of shell type with fire tube design so as to achieve high heat transfer rates by both radiation and convection (Refer Figure 4.4).

Fig. 4.4: Packaged Boilers

4.4.4. Stoker Fired Boiler

Stokers are classified according to the method of feeding fuel to the furnace and by the type of grate. The main classifications are:

1. Chain-grate or travelling-grate stoker
2. Spreader stoker

4.4.5. Chain-Grate or Travelling-Grate Stoker Boiler

Coal is fed onto one end of a moving steel chain grate. As grate moves along the length of the furnace, the coal burns before dropping off at the end as ash. Some degree of skill is required, particularly when setting up the grate, air dampers and baffles, to ensure clean combustion leaving minimum of unburnt carbon in the ash.

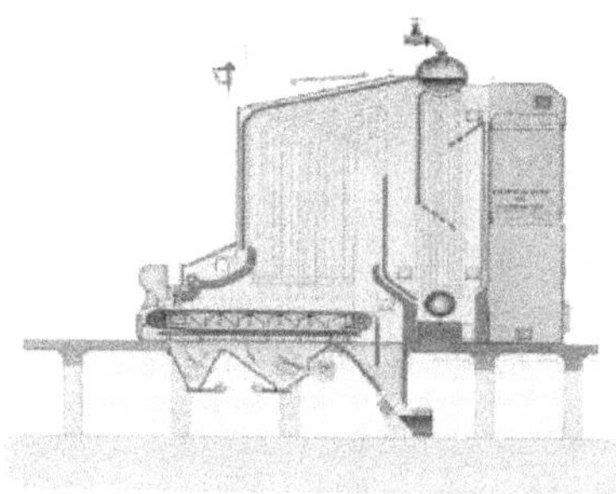

Fig. 4.5: Chain Grate Boiler

The coal-feed hopper runs along the entire coal feed end of the furnace. A coal grate is used to control the rate at which coal is fed into the furnace, and to control the thickness of the coal bed and speed of the grate. Coal must be uniform in size, as large lumps will not burn out completely by the time they reach the end of the grate. As the bed thickness decreases from coal-feed end to rear end, different amounts of air are required- more quantity at coal-feed end and less at rear end (see Figure 4.5).

4.4.6. Spreader Stoker Boiler

Spreader stokers utilize a combination of suspension burning and grate burning. The coal is continually fed into the furnace above a burning bed of coal. The coal fines are burned in suspension; the larger particles fall to the grate, where they are burned in a thin, fast-burning coal bed. This method of firing provides good flexibility to meet load fluctuations, since ignition is almost instantaneous when firing rate is increased. Hence, the spreader stoker is favoured over other types of stokers in many industrial applications.

4.4.7. Pulverized Fuel Boiler

Most coal-fired power station boilers use pulverized coal, and many of the larger industrial water-tube boilers also use this pulverized fuel. This technology is well developed, and there are thousands of units around the world, accounting for well over 90% of coal-fired capacity. The coal is ground (pulverised) to a fine powder, so that less than 2% is +300 micro metre (μm) and 70-75% is below 75 microns, for a bituminous coal. It should be noted that too fine a powder is wasteful of grinding mill power. On the other hand, too coarse a powder does not burn completely in the combustion chamber and results in higher unburnt losses. The pulverised coal is blown with part of the combustion air into the boiler plant through a series of burner nozzles. Secondary and tertiary air may also be added. Combustion takes place at temperatures from 1300-1700°C, depending largely on coal grade. Particle residence time in the boiler is typically 2 to 5 seconds, and the particles must be small enough for complete combustion to have taken place during this time. This system has many advantages such as ability to fire varying quality of coal, quick responses to changes in load, use of high pre-heat air temperatures etc. One of the most popular systems for firing pulverized coal is the tangential firing using four burners corner to corner to create a fireball at the centre of the furnace (see Figure 4.6).

4.4.8. FBC Boiler

When an evenly distributed air or gas is passed upward through a finely divided bed of solid particles such as sand supported on a fine mesh, the particles are undisturbed at low velocity. As air velocity is gradually increased, a stage is reached when the individual particles are suspended in the air stream. Further, increase in velocity gives rise to bubble formation, vigorous turbulence and rapid mixing and the bed is said to be fluidized.

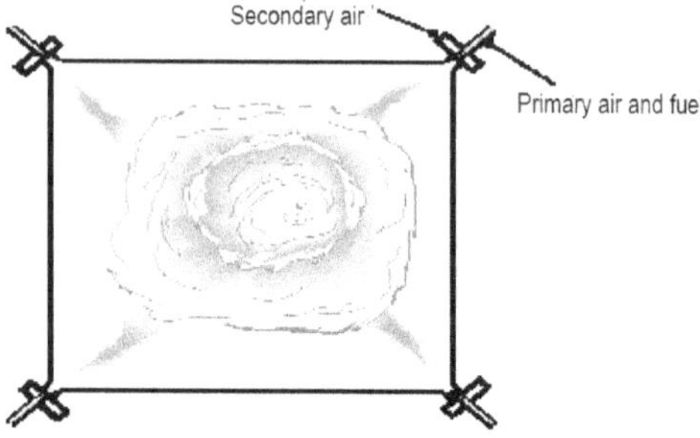

Fig. 4.6: Pulverized Fuel Boiler

If the sand in a fluidized state is heated to the ignition temperature of the coal and the coal is injected continuously in to the bed, the coal will burn rapidly, and the bed attains a uniform temperature due to effective mixing. Proper air distribution is vital for maintaining uniform fluidization across the bed.). Ash is disposed by dry and wet ash disposal systems. Fluidised bed combustion has significant advantages over conventional firing systems and offers multiple benefits namely fuel flexibility, reduced emission of noxious pollutants such as SOx and NOx, compact boiler design and higher combustion efficiency.

4.5. Boiler Types (based on live steam)

Live steam models utilize many different varieties of boilers ranging from the simple pot to the locomotive type. Each boiler type can give excellent performance so long as it is operated within its design envelope. Copper is the best material for small boilers. Brass should never be used for a boiler barrel, but is satisfactory for fittings.

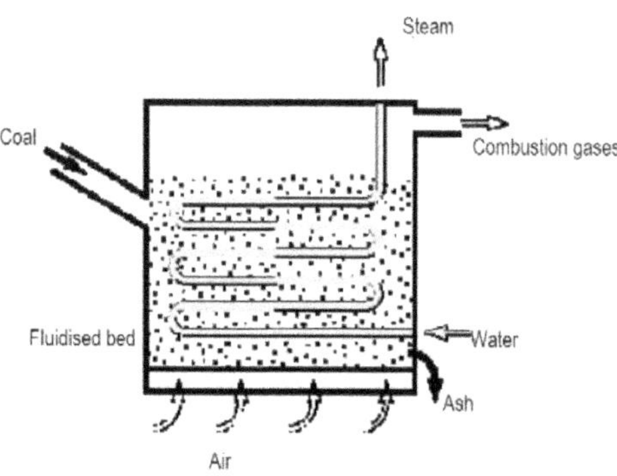

Fig. 4.7: FBC Boiler

Major Boiler Types are Discussed as Follows

1. Pot Type
2. Water Tube Type (Smithies)
3. Vertical Type
4. Centre Flue Type
5. Smoke Tube Type
6. Type C
7. Locomotive Type
8. Saddle Type

4.5.1. Pot Type

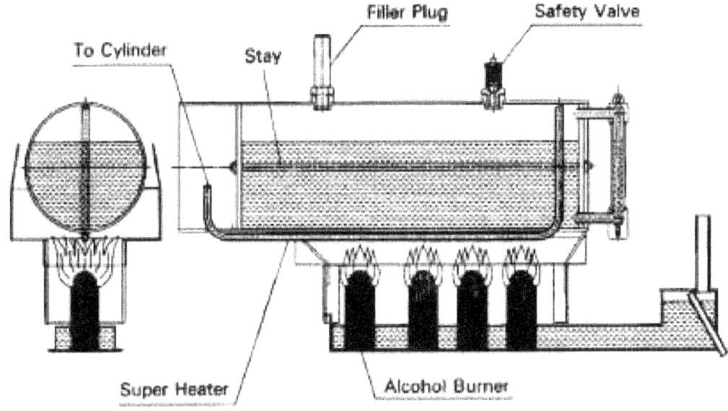

Fig. 4.8: Pot Type Boiler

The pot boiler, show in Figure 4.8, is the simplest type and consists of a cylindrical copper tube with stayed end plates. The fire, which is typically from an alcohol burner, is applied to the external surface of the boiler. Its steaming ability can be significantly increased by the addition of a smoke tube and a stainless steel shield which encloses the burner and the lower portion of the boiler. Thus configured, the pot boiler can be a god steam generator in moderate temperatures and mild winds.

Tank locomotives are good candidates for pot boilers since the tanks hide the fire shield. A suction fan and blower are not needed for firing since there is no necessity for a forced draft. Pot boilers have large water capacity and simple to steam. However, pot boilers tend to discolor since the burner flame is in direct contact with the outer barrel surface. Either a wick or vaporizing type burner can be used.

4.5.2. *Water Tube Type (Smithies)*

The water tube boiler was developed in the United Kingdom at the turn of the century by Mr. Fred Smithies. It consists of a copper inner barrel with stayed end plates and water tubes at the barrel's bottom, which extends from the front to the rear. This assembly slips into an outer casing which is shaped to resemble a locomotive boiler and firebox. There is an optimum sizing and arrangement for the water tubes, which depends on the boiler's dimensions.

Typically, the tubes should be greater than 5 mm in diameter and should be space so as to allow plenty of room for a flame path between them. Increasing the number of tubes may not increase the boiler's ability to generate steam. If the flame path is restricted, the boiler cannot perform well since the draft cannot draw the flame forward toward the smoke box and increase flame temperature. The inner surface of the outer casing is insulated with a ceramic sheet. It is important that the insulation does not obstruct the flame path near the throat plate. Properly designed Smithies boilers perform well in adverse weather conditions. If the boiler should run dry, the lack of draft will automatically extinguish the fire. A Smithies boiler is simple to build and operate, but it does require a suction fan and a blower since it is a forced draft system. The biggest disadvantage of the Smithies boiler is the limited water capacity of its inner barrel as compared to other boiler types with similar external dimensions.

4.5.3. *Vertical Type*

The vertical boiler is a simple type, which consists of a firebox at the bottom and a copper barrel with a smoke tube. It typically is used to drive stationary engines and boats. Firing is accomplished by alcohol or solid fuel pellets. More sophisticated versions of the vertical boiler contain many small tubes and are sometimes fired by coal or charcoal.

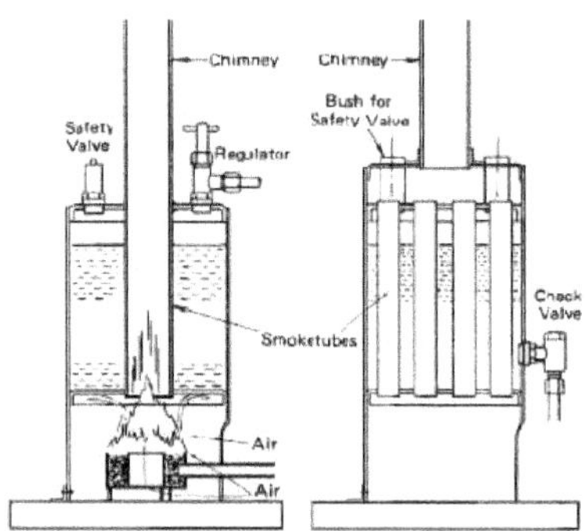

Fig. 4.9: Two Types of Vertical Boilers

4.5.4. *Center Flue Type*

The center flue boiler, show in Figure 4.10, has a large water capacity and a low center of gravity which makes it ideal for model boats. The center flue is surrounded by water and sometimes has several cross tubes to improve circulations. This type of boiler is usually fired by a gas burner, because the flame is completely enclosed by the center flue. Therefore, the probability of an accidental fire is reduced. It is necessary to maintain the proper water level in this type of boiler to avoid damaging the center flue. It offers good performance capabilities in adverse weather conditions.

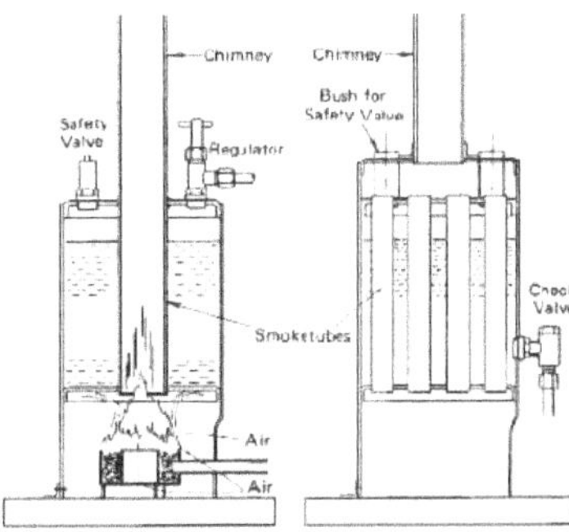

Fig. 4.10: Center Flue Type Boiler

4.5.5. *Smoke Tube Type*

The illustration in 4.11. Shows the Mixed Type of boiler developed by Dr. Seiichi Atonable. It has water tubes filled in the firebox of a locomotive type boiler in such as a way that the tubes act a stays. This configuration also increases the heating area and provides better water circulation than a plan locomotive type boiler. The Aster JNR D 51 and JNR 9600 utilize this type of boiler

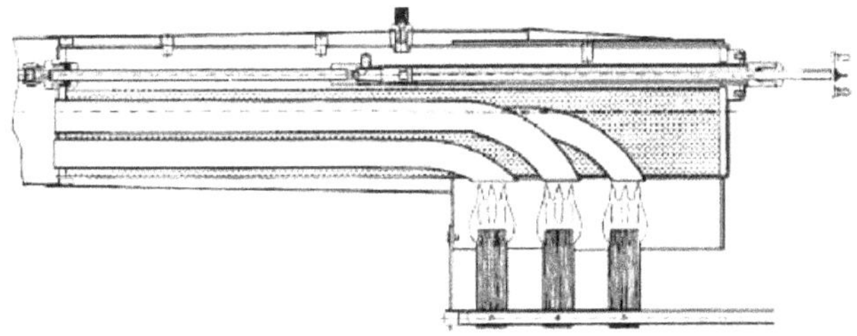

Fig. 4.11: Smoke Tube Type Boiler

4.5.6. *Locomotive Type*

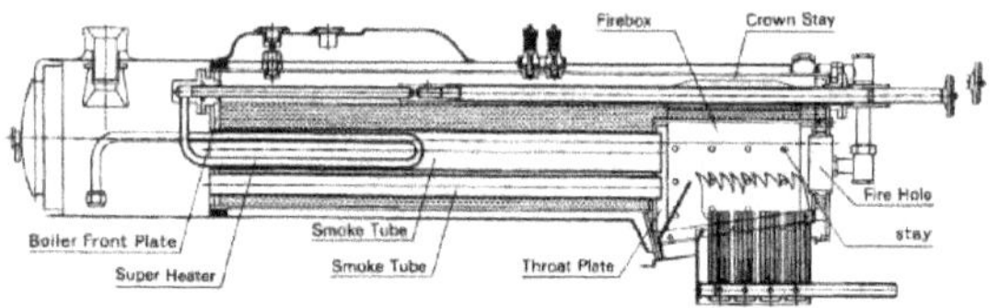

Fig. 4.12: Locomotive Type

The Locomotive type boiler, shown in Figure 4.12, consists of a copper barrel and an outer and inner firebox attached to the rear of the barrel. Perfectly round pressure vessels carry pressure loads by developing hoop tension stresses in the wall of the boiler. In this case, there is no tendency for the walls to bend. If a pressure vessel is not perfectly round, the walls will bend because the walls will bend attempting to form a perfect circle as it pressurizes. The results in a combined stress field with high stress levels. Since the firebox is not circular, it is necessary to provide additional structural members to strengthen non circular surfaces. This is the reason for stay bolts and structural beams in the firebox assembly.

Thermal stresses also occur as a result of thermal expansion, therefore a boiler must be designed to carry both thermal and pressure loads. In a Belpaire type firebox, the outer wrapper and crown sheet may be stiffened by beams or girder stayed to each other. Both side plates of the outer firebox are cross stayed since they are flat. Flat surfaces on the back head and smoke box tube plate can be stiffened or stayed. Sometimes a combination of both is used resulting in a very strong boiler. If two or more stays are used, once can be hollow and used to route the blower line to the smoke box. Tubes extend from the front tube plate in the smoke box to the rear tube plate, which is located at the front of the fire box.

4.5.7. *Saddle Type*

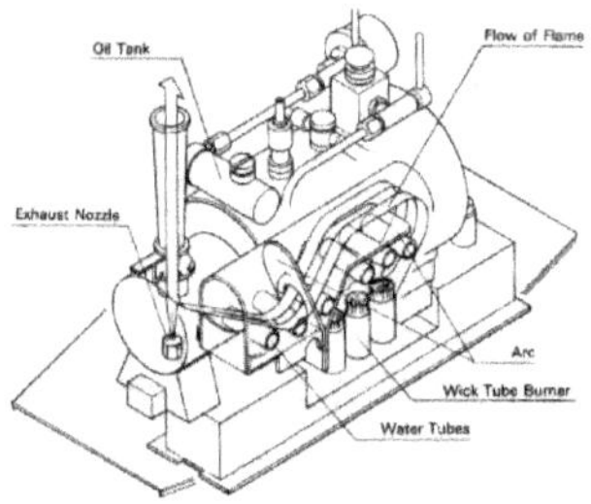

Fig. 4.13: Saddle Type Boiler

The saddle type boiler, as shown in Figure 4.13 is used for the Aster Glaskasten. In order to provide a reasonable running period, it was necessary to have a boiler with a minimum water capacity of 100 cc.

Since space was limited, a unique boiler shape was required. The saddle type boiler consists of a large diameter copper tube with end plates and a small cross tube. It uses features of the type C boiler as well as the Pot Type and Cross Water Tube Type. The inclined cross tube provides both extra water capacity and additional heating surface. They improve water circulation and also act as stays. The arc is filled a stainless steel firebox guiding the flame flow, so that the maximum heating area is realized.

This type of boiler can be fired by coal if grates are installed. Liquid fuel burners can also be employed. The locomotive type boiler is the most difficult to build but is very efficient and allows the realism of a coal fire.

It requires a suction fan and a blower, since it is a forced draft system. Its water capacity may be less than a comparable type C boiler. The water level must be carefully maintained so that the crown sheet is always covered; otherwise the boiler will be damaged.

4.6. Types of Boilers (Based on Industries Use)

Types of Boilers

Based on the order of Evaporation, the boilers are categorized into:

4.6.1. Cast Iron sectional Boilers

These are used for hot water services with a maximum operating pressure of 5 bar and a maximum output in the order of 1500 KW. Site assembly of the unit is necessary and will consist of a bank of cast iron sections. Each section has internal waterways. The sections are assembled with screwed or taper nipples at top and bottom for water circulation and sealing between the sections to contain the products of combustion. Tie rods compress the sections together.

A standard section may be used to give a range of outputs dependent on the number of sections used. After assembly of the sections, the mountings, insulation and combustion appliance are fitted. This system makes them suitable for locations where it is impractical to deliver a package unit, e.g. basements where inadequate access is available or rooftop plant rooms where sections may be taken up using the elevator shafts. Models available use liquid, gaseous and solid fuel.

4.6.2. Steel Boilers

These are similar in rated outputs to the cast iron sectional boiler. Construction is of rolled steel annular drums for the pressure vessel. They may be of either vertical or horizontal configuration, depending upon the manufacturer. In their vertical pattern they may be supplied for steam rising.

4.6.3. Electrode Boilers

These are available for steam rising up to 3600kg/h and manufacture is of two designs. The smaller units are element boilers with evaporation less than 500kg/h. In these, an immersed electric element heats the water and a set of water-level probes positioned above the element controls the water level being interconnected to the feed water pump and the element electrical supply.

Larger units are electrode boilers. Normal working pressure would be 10 bar but higher pressures are available. Construction is a vertical pattern pressure shell containing the electrodes. The lengths of the electrodes control the maximum and minimum water level. The electrical resistance of the water allows a current to flow through the water, which in turn, boils and releases steam. Since water has to be present within the electrode system, lack of water cannot burn out the boiler. The main advantage with these units is that they may be located at the point where steam is required and, as no combustion fumes are produced, no chimney is required. Steam may also be raised relatively quickly, as there is little thermal stressing to consider.

4.6.4. Steam Generators

This coil type boilers works in the evaporative range up to 3600 kg/h of steam. Because of the steam pressure being contained within the tubular coil, pressures of 35bar and above are available, although the majority is supplied to operate at up to 10 bars. They are suitable for firing with liquid and gaseous fuels, although the use of heavy fuel oil is unusual. The coiled tube is contained within a pressurized combustion chamber and receives both radiant and convected heat. A control system matches the burner-firing rate proportional to the steam demand. Feed water is pumped through the coil and partially flashed to steam in a separator. The remaining water is recirculated to a feed water heat exchanger before being run to waste. Because there is no stored water in this type of unit they are lighter in weight and therefore suitable for sitting on mezzanine or upper floors adjacent to the plant requiring steam. Also, as the water content is minimal, steam raising can be achieved very quickly and can respond to fluctuating demand within the capacity of the generator. It must be noted that close control of suitable water treatment is essential to protect the coil against any build-up of deposits.

4.6.5. Vertical Shell Boilers

This is a cylindrical boiler where the shell axis is vertical to the firing floor. Originally it comprised a chamber at the lower end of the shell, which contained the combustion appliance. The gases rose vertically through a flue surrounded by water. Large diameter (100mm) cross tubes were fitted across this flue to help extract heat from the gases which then proceeded to the chimney. Later versions had the vertical flue replaced by one or two banks of small-bore tubes running horizontally before the gases discharged to the chimney. The steam was contained in a hemispherical chamber forming the top of the shell. The present vertical boiler is generally used for heat recovery from exhaust gases from power generation or marine applications. The gases pass through small-bore vertical tube banks. The same shell may also contain an independently fired section to produce steam at such times, as there is insufficient or no exhaust gas available.

4.6.6. Waste Heat Boilers

These may be horizontal or vertical shell boilers or water tube boilers. They would be designed to suit individual applications ranging through gases from furnaces, incinerators gas turbines and diesel exhausts. The prime requirement is that the waste gases must contain sufficient usable heat to produce steam or hot water at the condition required. Supplementary firing equipment may also be included if a standby heat load is to be met and the waste-gas source is intermittent. Waste-heat boilers may be designed to us either radiant or convected heat sources. In some cases, problems may arise due to the source of waste heat, and due consideration must be taken of this, with examples being plastic content in waste being burned in incinerators, carry-over from some type of furnaces causing strongly bonded deposits and carbon from heavy oil fired engines. Some may be dealt with by maintaining gas-exit temperatures at a predetermined level to prevent dew point being reached and others by soot blowing. Currently, there is a strong interest in small combined heat and power (CHP) stations, and these will normally incorporate a waste-heat boiler.

4.6.7. Fluid-bed Boilers

The name derives from the fire bed produced by containing a mixture of silica sand and ash through which air is blown to maintain the particles in suspension.

The beds are in three categories

i)   Shallow bed

ii) Deep bed and

iii) Recirculation bed

Shallow beds are mostly used and are about 150-250 mm in depth in their slumped condition and around twice that when fluidized. Heat is applied to this bed to raise its temperature to around 600C by auxiliary oil or gas burners. At this temperature coal and/or waste is fed into the bed, which is controlled to operate at 800-900C.Water-cooling surfaces are incorporated into this bed connected to the water system of the boiler. The deep bed, as its name implies, is similar to the shallow bed but in this case may be up to 3m deep in its fluidized state, making it suitable only for large boilers. Similarly, the recirculating fluid bed is only applicable to large water tube boilers.

Several applications of the shallow-bed system are available for industrial boilers; the two most used being the open-bottom shell boiler and the composite boiler. With the open-bottom shell the combustor is sited below the shell and the gases then pass through two banks of horizontal tubes. In the composite boiler the combustion space a water tube chamber directly connected to a single-pass shell boiler forms housing the fluid bed.

In order to fluidize the bed the fan power required will be greater than that with other forms of firing equipment. To its advantage, the fluid bed may utilize fuels with high ash contents, which affect the availability of other systems. It is also possible to control the acid emissions by additions to the bed during combustion. They are also less selective in fuels and can cope with a wide range of solid-fuel characteristics.

4.7. Types of Boilers (Auxiliary Type)

Auxiliary boilers may be divided into two groups: FIRE-TUBE BOILERS and WATER-TUBE BOILERS.

4.7.1. Fire-Tube Boilers

Fire-tube boilers are generally similar to Scotch marine or locomotive boilers. In this type of boiler, the gases of combustion pass through tubes that are surrounded by water. There are a number of auxiliary boilers of the fire-tube type in use in diesel-driven ships. Figure illustrates a cutaway view of the fire-tube boiler shown in figure.

4.7.2. Water-Tube, Natural-Circulation Boilers

Water-tube, natural-circulation boilers consist basically of a steam drum and a water drum connected by a bank of generating tubes. The two drums are also connected by a row of water tubes, which forms a water-cooled sidewall opposite the tube bank. The water-wall tubes pass beneath the refractory furnace floor before they enter the water drum.

In natural-circulation boilers, the steam and water drums are connected by several tubes of larger diameter, called DOWNCOMERS or WATER TUBES (not shown). These tubes are positioned away from the flow of hot gases of combustion. Refractory is also used to protect these down comers from contact with the combustion gases.

The operating principle of a natural-circulation boiler is quite simple. It relies on the difference in density (weight) between the cooler (heavier) water in the water tubes (or down comers) and the hot, less dense (lighter) water in the steam-generating tubes. This is the force that causes the hot water and steam mixture to rise in the tubes in the generating bank, from the water drum to the steam drum, where the steam is separated from the water and rises to the top of the steam drum. The flow of water up the tubes of the steam-generating bank must be maintained; otherwise, the tubes would quickly melt. A constant flow of water and steam up the tubes is required to carry away heat at the proper rate. If the flow from natural circulation is allowed to stop, such as when the water level in the steam drum falls below the openings of the bank of tubes for the water wall, the tubes of the generating bank will be severely damaged and the boiler will need major repairs. (Replacing boiler tubes is an expensive operation.)

4.8. Super Heaters

One of the most important accessories of a boiler is a super heater.

It affects improvement and economy in the following ways.

- The super heater increases the capacity of the plant.
- Eliminates corrosion of the steam turbine.

Reduces steam consumption of the steam turbine.

Super heater coils, most of them are very complicated in shape, gauge and size. In order to resist metal temperatures above 600°C different types of materials were used for the tubes. The materials range from M.S., seamless carbon steel tubes to chromium molybdenum seamless alloy steel tubes to Stainless Steel Tube.

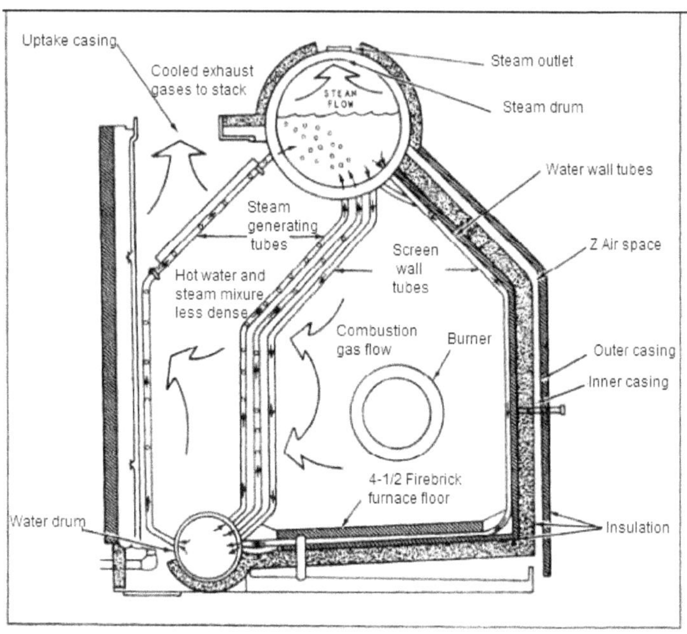

Fig. 4.14: Natural-circulation, Water-tube Boiler

Tube weld earned remarkable achievement by applying creativity used in choosing extra-ordinary materials after a bit of hectic experiments and with varying thickness depending upon the design and system aspect, the location, the relative temperature prone area etc. for boilers up to 210 MW power plant.

4.9. Economisers

In case of boiler system, economizer can be provided to utilize the flue gas heat for pre-heating the boiler feed water. On the other hand, in an air pre-heater, the waste heat is used to heat combustion air. In both the cases, there is a corresponding reduction in the fuel requirements of the boiler. A economizer is shown in Figure 8.8. For every 220 C reduction in flue gas temperature by passing through an economizer or a pre-heater, there is 1% saving of fuel in the boiler. In other words, for every 60 C rises in feed water temperature through an economiser, or 200C rise in combustion air temperature through an air pre-heater, there is 1% saving of fuel in the boiler. These are feed-water heaters in which the heat from waste gases is recovered to raise the temperature of feed-water supplied to the boiler.

They offer the following advantages:

- Fuel economy
- Longer life of the boiler
- Increase in steaming capacity
- Finned Tube Economisers
- C.I. Gilled Tube Economisers
- Plain Tube Coil Economisers

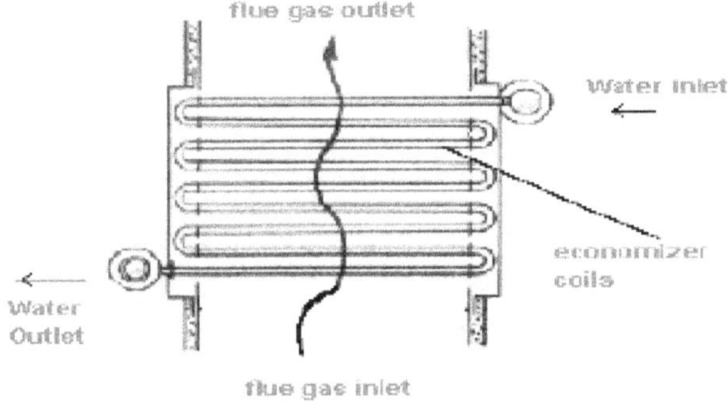

Fig. 4.15: Economisers

4.10. Air Heater (or) Recuperators

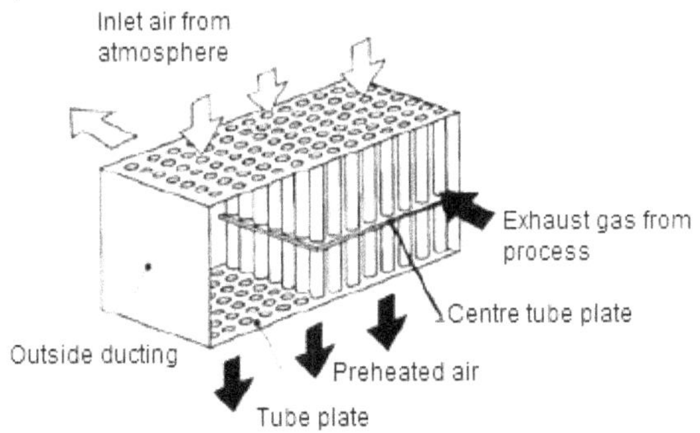

Fig. 4.16: Air Heater

In a recuperator, heat exchange takes place between the flue gases and the air through metallic or ceramic walls. Duct or tubes carry the air for combustion to be preheated; the other side contains the waste heat stream. A recuperator for recovering waste heat from flue gases is shown in Figure 4.16. The simplest configuration for a recuperator is the metallic radiation recuperator, which consists of two concentric lengths of metal tubing as shown in Figure 4.17.

The inner tube carries the hot exhaust gases while the external annulus carries the combustion air from the atmosphere to the air inlets of the furnace burners. The hot gases are cooled by the incoming combustion air which now carries additional energy into the combustion chamber. This is energy which does not have to be supplied by the fuel; consequently, less fuel is burned for a given furnace loading. The saving in fuel also means a decrease in combustion air and therefore stack losses are decreased not only by lowering the stack gas temperatures but also by discharging smaller quantities of exhaust gas.

The radiation recuperator gets its name from the fact that a substantial portion of the heat transfer from the hot gases to the surface of the inner tube takes place by radioactive heat transfer. The cold air in the annuals, however, is almost transparent to infrared radiation so that only convection heat transfer takes place to the incoming air. As shown in the diagram, the two gas flows are usually parallel, although the configuration would be simpler and the heat transfer more efficient if the flows were opposed in direction (or counter flow). The reason for the use of parallel flow is that recuperators frequently serve the additional function of cooling the duct carrying away the exhaust gases and consequently extending its service life.

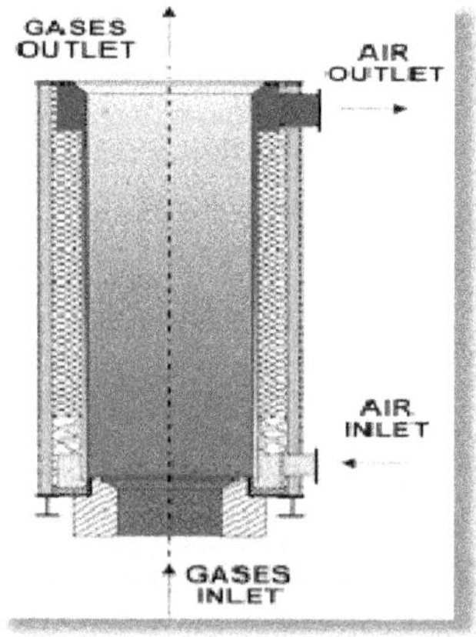

Fig. 4.17: Air Heater (or) Recuperators

4.11. Water Wall Tubes

These are tubes in the Boiler where water is evaporated to steam and are also called Steam Generating Tubes. These Tubes also form the Walls of the Boiler and are hence called Water Walls or Water Wall Panels. These Tubes have very complicated shapes to allow Inspection openings and burner throats and fabrication require intricate binding on CNC programmable bending Machines and checking on 3D layouts.

4.12. Boiler Blow down

When water is boiled and steam is generated, any dissolved solids contained in the water remain in the boiler. If more solids are put in with the feed water, they will concentrate and may eventually reach a level where their solubility in the water is exceeded and they deposit from the solution. Above a certain level of concentration, these solids encourage foaming and cause carryover of water into the steam. The deposits also lead to scale formation inside the boiler, resulting in localized overheating and finally causing boiler tube failure. It is, therefore, necessary to control the level of concentration of the solids and this is achieved by the process of 'blowing down', where a certain volume of water is blown off and is automatically replaced by feed water-thus maintaining the optimum level of total dissolved solids (TDS) in the boiler water. Blow down is necessary to protect the surfaces of the heat exchanger in the boiler. However, blow down can be a significant source of heat loss, if improperly carried out. The maximum amount of total dissolved solids (TDS) concentration permissible in various types of boilers is given in Table 4.1.

Table 4.1: The Maximum Amount of Total Dissolved Solids (TDS) Concentration Permissible in Various Types of Boilers

Boiler Type	Maximum TDS (ppm)*
Lancashire	10,000 ppm
Smoke and water tube boilers (12 kg/cm^2)	5,000 ppm
Low pressure Water tube boiler	2000-3000
High Pressure Water tube boiler with super heater etc	3,000 - 3,500 ppm
Package and economic boilers	3,000 ppm
Coil boilers and steam generators	2000 (in the feed water)

4.13. Intermittent Blow Down

The intermittent blown down is given by manually operating a valve fitted to discharge pipe at the lowest point of boiler shell to reduce parameters (TDS or conductivity, pH, Silica and Phosphates concentration) within prescribed limits so that steam quality is not likely to be affected. In intermittent blow down, a large diameter line is opened for a short period of time, the time being based on a thumb rule such as "once in a shift for 2 minutes". Intermittent blow down requires *large* short-term increases in the amount of feed water put into the boiler, and hence may necessitate larger feed water pumps than if continuous blow down is used. Also, TDS level will be varying, thereby causing fluctuations of the water level in the boiler due to changes in steam bubble size and distribution which accompany changes in concentration of solids. Also substantial amount of heat energy is lost with intermittent blow down.

4.14. Continuous Blow Down

There is a steady and constant dispatch of small stream of concentrated boiler water, and replacement by steady and constant inflow of feed water. This ensures constant TDS and steam purity at given steam load. Once blow down valve is set for a given conditions, there is no need for regular operator intervention. Even though large quantities of heat are wasted, opportunity exists for recovering this heat by blowing into a flash tank and generating flash steam. This flash steam can be used for pre-heating boiler feed water or for any other purpose (see Figure 4.18 for blow down heat recovery system). This type of blow down is common in high-pressure boilers.

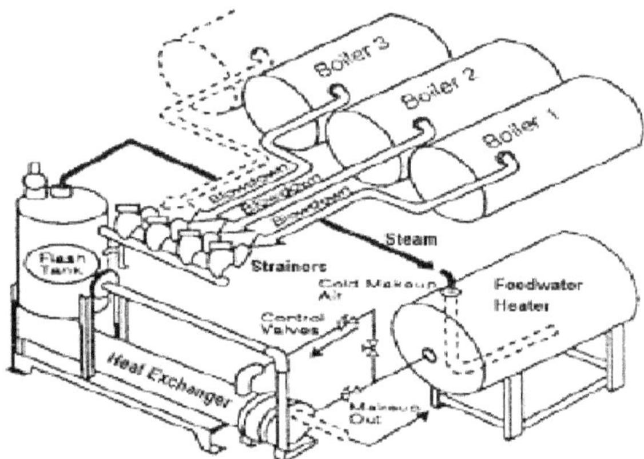

Fig. 4.18: Blow Down Heat Recovery System

4.15. Boiler Water Treatment

Any steam boiler can be expected to have a considerable life period provided that it is properly looked after. One of these basic maintenance requirements is the careful control of the boiler water chemistry. World wide experience has proved the benefits of a chemically clean boiler, and a locomotive owner can do much to prolong the life of the boiler by keeping it thoroughly clean both internally and externally and not subjecting it to rapid or frequent changes in temperature.

The water that is carried on the locomotive, the Feed Water, is generally supplied from the towns' mains. This is essentially fresh water, however, it contains many impurities, and when fed into the boiler via the injector, these impurities become a significant problem. It is during the process of evaporation that these impurities are deposited on the heating surfaces as scale, remain in solution to attack the metal plates chemically, or are carried to the surface causing the effects known as Priming and Foaming.

The quality of towns' mains water will vary according to location around the country, and it is sometimes termed Soft or Hard. Soft water contains much fewer impurities and so has less scale forming salts, however it is usually of an acid nature and this is damaging to the metal plates causing corrosion. Hard water is what we find in the southern region, this has many scale forming salts but is generally of an Alkaline nature and so protects the boiler plates from acid attack.

Hard water contains, Calcium and Magnesium Bicarbonates, and calcium carbonate, these are the principle soft fur scales that adhere to heating surfaces. Calcium and Magnesium Sulphate are also present and these are the principle Hard forming scales which become baked onto the heating surfaces and prove very difficult to remove once established. Scale on the heating surfaces is an excellent insulator of heat and a coating as little as 1/8" of an inch will increase the fuel consumption by as much as 20%. Large or thick coatings of scale will cause overheating of the firebox plates and these can then burn away causing thinning and buckling of the plates, leaking tubes and leaking stays. Oil on the heating surfaces is another excellent insulator and a coating of only a few thousands of an inch will be equivalent to about half an inch of scale.

Do not be confused between boiler water additive and oils like diesel; these are often stored in similar drums!

Chemically softening hard water by passing it through a water softener will exchange the scale forming, (Calcium and Magnesium) salts into more soluble Sodium salts, (sodium bicarbonate) which decomposes to carbon dioxide when heated. These salts will remain in the boiler water an add to what is known as the Total Dissolved Solids content, or TDS Level. This becomes more concentrated as the boiler evaporates more water.

4.16. Scale Prevention

In order to keep the boiler as free of scale as possible it is necessary to treat it. This could be done by feeding it Softened Water, but this can be costly. The other method is to treat it chemically. This is done by adding a premixed chemical compound to the feed water in the tender tank. Both methods using Softened water and Chemicals can be used together with excellent results, but do require strict blowing down routines.

This chemical compound that is currently used is supplied by Freeston and Sons and is called "L4". It is very dark brown/red in colour and has little or no smell. It contains a complex blend of Phosphates, Tannins and starch, these are designed to convert the calcium and magnesium salts to insoluble compounds and so form a chemical sludge before they can be deposited on the heating surfaces. This sludge falls to the bottom of the boiler and is removed periodically by blowing down and washing out.

4.17. Corrosion Prevention

The boiler structure must be protected from the effects of chemical attack which will cause corrosion and eventual failure of its component parts. The feed water must be treated to make sure that chemical corrosion attack cannot take place. (see diagram. L) The ideal conditions for boiler water is mildly Alkaline, this then being able to mop up and neutralize any acids that that are fed into it. Sodium Hydroxide is a strong Alkali and a weak solution of this is blended to the boiler water compound and all though the water supply is alkaline in nature, (hard water areas only), the Sodium Hydroxide helps to keep the boiler alkalinity topped up. Soft water areas are by nature slightly acid and so require more of this chemical in its water treatment.

4.18. Oxygen Attack

Apart from the scale forming salts, the feed water is usually rich in dissolved Oxygen. Iron contained in the steel plates of the boiler will react chemically with oxygen causing rapid corrosion, this is accelerated as the temperature and pressure gets higher. To prevent Oxygen attack Oxygen must be removed from the water, and this can be done chemically by the addition of an Oxygen Scavenger. This is blended into the boiler water compound. Sodium Sulphite is the commonest of these and as it reacts with the oxygen it forms a soluble non scale forming salt and remains in the water adding to the Total Dissolved Solids level.

4.19. Total Dissolved Solids

The addition of feed water, with its dissolved salts, scale prevention, anticorrosion and oxygen scavenging chemicals is being constantly added to the boiler and the pure water is being evaporated away, these chemicals and salts remain in the water, concentrating all the time. This is the Total Dissolved Solids content.

4.20. Priming and Foaming

Continued increase in the TDS levels may reach a concentration where another condition known as FOAMING may result. The effects of Foaming can be serious as the water will foam up in the boiler and be carried over with the steam depositing the salts in the super heater and causing the water to pass into the cylinders causing serious damage. Blowing down will reduce the effects of foaming by removing the dissolved salts from the water. Continued Priming and Foaming is an indication the boiler is in need of a washout. The boiler water compound contains a small quantity of a chemical called Polyamides or Antifoams, this reduces the film around the steam bubbles causing them to collapse more readily and so helping to prevent foaming occurring. In cases of severe contamination separate Antifoam compound may be added in addition to the boiler water compound.

PRIMING, which is the carryover of boiler water to the cylinders, is likely to occur when the TDS levels are high making the water more unstable and is more likely to occur running with a high water level, and a heavy steam demand allowing the water to easily be carried over.

OIL contamination in the water will cause severe priming and all steps to prevent oil entering the boiler water spaces and feed water must be taken.

4.21. Burners and Controls

4.21.1. Atmospheric (Venturi) Burners

These are burners that operate on gas only. The velocity of the gas stream flowing through an orifice entrains atmospheric air for combustion from a venturi throat. The resultant mixture burns at a specially designed tip, of which there are a wide variety, known as a flame retention head. For hot glass glory-hole, tank and pot furnaces, atmospheric burners generally use L.P. Gas at high pressure as the pressure of reticulated natural gas is usually too low to inspirate sufficient air to generate a hot, short flame with any forward velocity. Atmospheric natural gas burners are used successfully on ceramic kilns.

Fig. 4.19: Atmospheric Burners

There is a definite ratio between the burner port area and the venturi throat. Typically, depending on the kiln or furnace back pressure, gas pressure and burner head design, the venturi throat area should be approx. 40-50% of the total burner port area as shown in figure 4.19. Mismatching may result in a decrease in mixture velocity producing inadequate burning or at worst "flash back". Low rates can be achieved by using a preheat pilot.

This method of firing is the cheapest in terms of equipment costs but is also the least economical. Attention should be paid to obtaining as neutral (correct mixture of atmospheric air and gas) a flame as possible as either oxidizing (more air then gas) or reducing (less air) flames or atmospheres are wasteful. Special process or glaze firings are of course excluded. Although some secondary air (air entrained around the burner tip) is necessary for complete combustion, try to keep it to a minimum by correctly sized burner ports and flue outlets. It is possible to check the furnace conditions by restricting the flue exit and checking for slight reduction. This will indicate the settings are close to perfect.

It is difficult to achieve sufficient temperature in higher temperature furnaces with these burners. The greatest limitation, due to the low mixture pressure produced, is the volume of combustion products that can be introduced to the furnace combustion area. Simply increasing the combustion space size will not achieve results as losses will increase proportionally. Obviously there is a fine balance. In some cases better results have been attained by using several smaller burners rather than a single, large burner. Smaller burners with smaller gas orifices and higher gas pressure would develop higher mixture pressures allowing more combustion products into the space in a shorter time. A better alternative is to increase the pressure the combustion products are forced into the space by using a forced air supply. Please contact for more advice on this option.

Flame safety systems for these burners are usually the thermoelectric type that can be used with or without a separate pilot burner. The pilot burner can serve as a low-fire setting. Mount the pilot and/or safety probe well away from the furnace back heat. It is always better and safer to purchase the burners assembled and pre-tested with the appropriate controls.

Automatic ignition and quick lockout safety systems are readily available as an option for these burners.

4.21.2. Pre-Mix Open and Sealed Burners

Fig. 4.20: Pre-Mix Open and Sealed Burners

These are burners using a machined mixing set and forced air from a blower or compressor. Open burners use a suitable cast iron or steel flame retention tip and the sealed type utilize a R.I. cast able tunnel or MP multiport tip (illustrated in fig. 4.20) mounted into the furnace wall. Natural gas or L. P. Gas may be used at low pressure as the forced air induces the gas and produces a blast-type flame. These burners are more efficient than atmospheric burners as greater control is available over the air and gas mix, a hotter flame is produced and the sealed burner requires no wasteful secondary air. The open burner will need secondary air for cooling purposes to prolong the tip's life and to complete combustion. Some operators have traded tip life for lower noise and lower capacity by sealing the tip in the furnace port with fibre.

The higher mixture pressures developed by this burner style enable greater combustion volumes into the available furnace combustion space. Some careful preheating of the combustion air may be possible but is not recommended, as there is a risk of flashback or damage to the controls through heat conduction.

Premix burners generally have a shorter flame length than comparable nozzle mix burners, the air/gas ratio is easier to control and the burner overall generally easier to set up. Special flow regulators can be fitted to simplify adjustment enabling alteration of air flow only to raise or lower the temperature. This lends itself to simple, accurate temperature control. Again, care must be taken to correctly size the mixer as a definite ratio exists between the size of the mixer chambers and the burner orifice for proper operation. It is possible to construct a simple mixer from pipe pieces but extra care should be taken to ensure it is not possible for the air to flow into the gas line if the burner or feed pipe is blocked. As a minimum, a light flap safety check valve should be fitted to the gas line. If accurate mixing and turn down are required, use the correct mixer.

State gas regulations concerning forced draft (premix, nozzle mix) burners tend to differ, with some States demanding full sequence electronic flame failure while others may allow glory hole burners without safety if the burner is constantly supervised. All enclosed kilns, pot or tank furnaces should have quick lock out safety as these burners can produce large amounts of unignited mixture in a short period of time.

4.21.3. Nozzle-Mix Burners

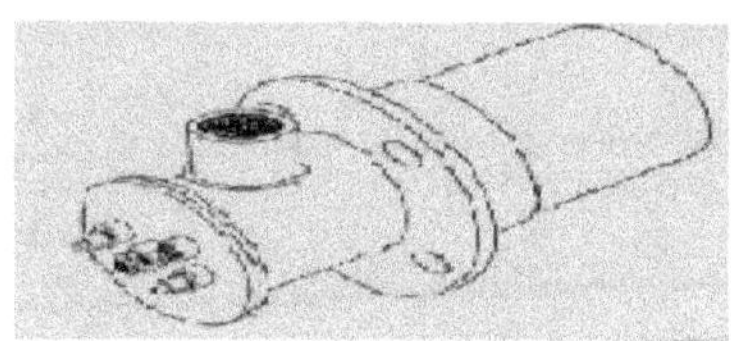

Fig. 4.21: Nozzle–mix Burners

These types of burners accomplish the mixing of the air and gas after they leave the burner port. Up to the burner head the air and gas are kept separate, lower gas and air pressures (unless high velocity is required) may be used and there is no chance of flash back. They generally have greater turn down than other burner types by controlling the gas only and can use preheated air. As the preheated combustion air is kept separate there is no chance of over-heating the gas controls.

Basic cast nozzle-mix burners can handle low preheats temperatures providing some fuel savings but deterioration may result if these burners are exposed to high temperatures for a period of time. Nozzle mix burners that will provide high preheat temperatures and maximum fuel savings are constructed from stainless steel internals or in some special cases ceramic materials. Various types of flame shapes and capacities can be designed to suit the customers' requirement including flat flame burners. This type of burner has a specially designed burner tip and refractory quarl to produce a spinning flat flame that spreads at 90 degrees to the mixture outlet. They have been used in industry where little forward flame travel is desirable and efficient radiant heat is best.

Air and gas controls and safety equipment are similar to premix burners.

4.22. Recuperation

Recuperation is the process of preheating the combustion air by utilizing the waste flue products. Although recuperation has been used in industry in various forms for many years, its use in small production situations has only recently become viable due to high fuel costs. Fuel savings of up to 40% can be achieved through properly designed and implemented systems. A simple counter flow design is illustrated in figure 4.22.

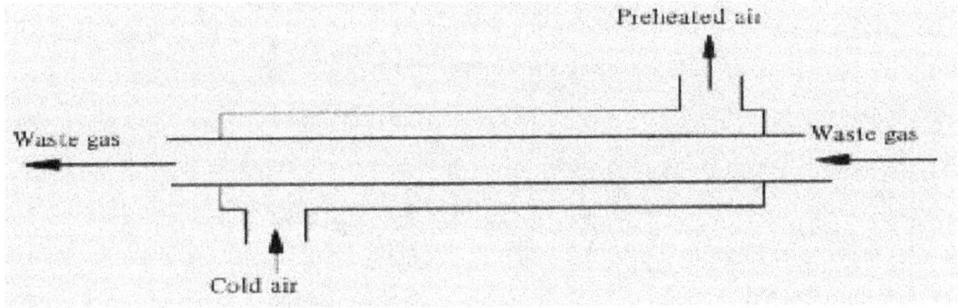

Fig. 4.22: Recuperation

A simple and effective recuperator uses a stainless steel tube on the flue outlet with a larger tube sited around it. The area between the inner and outer tubes must be adequate to allow for free circulation of the cold combustion air but sufficiently restricted to enable the heat conducted through the inner tube to heat the air to the desired preheat temperature by the time the air exits the recuperator and flows to the burner head. The passage of the hot combustion products flowing through the inner tube must similarly be restricted to allow sufficient heat penetration. Baffles may be added to the inner tube or the single inner tube exchanged for multiple tubes to increase the available surface area. It is important not to exceed the working temperature of the steel and a refractory base or longer section may need to be used to take the initial heat. Experience has shown that over a period of time a coating of products produced by the glass making process tend to accumulate on the inner tube surface. Provision must be made for periodic cleaning of the inside tube and it is a good idea to have a small reservoir underneath the flue outlet to gather these products rather than allow them to block the furnace outlet. Refractory recuperators can be made to increase the preheated air temperature to the burner and prolong life. Air piping from the recuperator to the burner head should have a large cross sectional area to cope with the expansion of the cold air as it is heated. This can be up to 40% at typical temperatures.

4.23. Firing Low Temperature Ovens

There are many functional ovens in use including some with well-designed features such as adjustable combustion spaces for varied throughput's and automatic doors for hands free loading. An efficient oven will have many capabilities including even heat distribution, the ability to maintain a set temperature and fire down over a period of time at an accurate rate if required as shown in figure 4.23.

Fig. 23: Low Temperature Ovens

Gas ovens work best using a down draft design (the flue outlet near the base of the oven) and burners that have a short, clean flame with good turndown characteristics (high flame to low flame). The burners usually fire through the base, on either side of the loading area, however higher gas pressure burners can fire horizontally along a base channel. Better ovens have been constructed using high pressure burners firing around the top of the space, creating a circular swirl, for even temperature gradients.

Insulation materials are a matter for personal preference with many people maintaining that R.I. brick ovens can virtually cools down (with all openings closed) at the required rate without any added heat input. Against this it must be remembered that it takes extra energy to heat a brick oven than a light-weight fibre type.

Low gas pressure burners suitable for these ovens are atmospheric and can be pipe-type burners with either a row of drilled holes, slots or newer designs that incorporate a mixer with many fine slots in the burner casing. The newer types are generally cheaper, more efficient and have a better turndown. Safety controls are usually thermoelectric.

As temperature control is such an important consideration, programmable units are available to accurately maintain and control the temperature gradient. Accurate electronic digital units are available up to 8 or 12 stages and control the burners using solenoid valves according to demand. Electric kilns are easily controlled with programmers, the contactor coil or relay substituted for the solenoid valve.

4.24. Higher Temperature Furnaces

These furnaces fire to higher temperatures (more than 700ºC) but can employ similar burners. Downdraft designs are effective for general work but it is also possible to fire high pressure smaller burners across the product. This method relies not only on convection heat input but also, to a degree, on radiation from the flame. It is important if using this method that sufficient draw is available from the fluing system particularly in the early stages of a firing. It may be necessary to preheat the flue but in any case pressure drops across the flue pipe and exit port must be minimized. Sealed burner tips are better. Larger, more industrial kilns should use higher pressure forced draft burners firing across, under or through the load. These use an air fan/blower with the appropriate mixing mechanism to deliver sufficient velocity for even heat distribution. Automatic temperature programmers are important and can be effectively connected with any gas burner system. Multiple tips commonly utilize a ladder type pilot arrangement on either side, modulating the main burners to relight from the pilot for accurate temperature control. Flame safety is simply fitted to the ladder pilot with manual or automatic spark ignition an option.

4.25. Safety Controls

Safety controls for gas appliances vary from over-temperature controls to sophisticated automatic start-up and flame failure controllers. Flame safety controls are recommended for all types of burners where there is the risk of a build-up of unignited gas should the flame are extinguished. Flame rectification or UV units are required on all forced air burners and gas will not be connected unless they are fitted. The regulations cover natural gas and L.P. Gas. In some instances the authorities may also insist on over temperature protection or explosion relief. Explosion relief is simply a panel fitted to the furnace of the correct dimensions of a material with less resistance to an explosion than the furnace itself. Non-return check valves may also be required in the gas main to prevent air flowing back to the meter. A regulator is always required on either natural gas or L.P. gas installations to monitor the gas flow and provide the correct pressure outlet. All valves and fittings must be approved and pipe fittings particularly must be suitable for gas. Safety valves should be listed in the gas association bulletin and the burners and controls installed by a licensed person.

4.26. Thermoelectric Safety

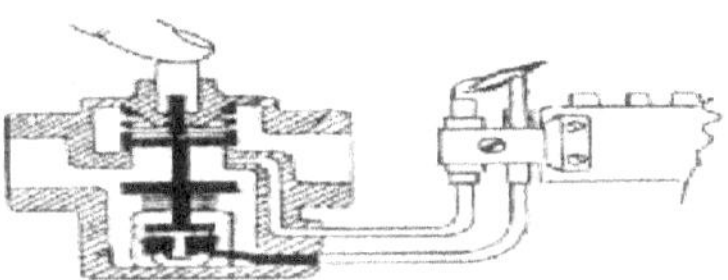

Fig. 4.24: Thermoelectric Safety

This is the simplest form of flame safety and is permitted on atmospheric burners with a capacity under 500 MJ/hour as shown in figure 4.24. These operate on an electromagnetic principle and require no power. A small current is generated when the tip of a thermocouple probe is heated by the flame. This current excites an electro-magnet located in a safety valve and attracts a plate allowing gas to flow. Shut off time, should the probe cool, can be up to 20 seconds.

4.27. Electronic Quick-Lockout

Fig. 4.25: Electronic Quick-layout

These units require power and are usually fitted to forced air burners or safer atmospheric burners. They shut down on a flame failure in approx. one second by closing a solenoid valve fitted into the gas line. Two main types are available; the flame rectification type and the ultra-violet type as shown in figure 4.25.. Flame rectification relies on the ability of ionized gases in the flame to rectify on AC current from the control unit. A flame sensing rod made from special high temperature material is used and must be situated near the edge of the main flame. A micro-amp meter may be used in series with the rod to check the best position and minimize nuisance shutdowns. The wire must be sturdy enough to resist drooping or deterioration at high temperatures and the porcelain insulators must be kept clean or replaced if cracks develop. For reliable operation the earthing point on the burner must have at least four times the area of the rod in the flame.

UV monitors are sensitive to the ultra violet radiation produced by flames. They also sense the arc of a spark so must be sited away from any automatic spark igniters fitted to the burner. Their main advantage in high temperature situations compared to flame rods is that they can be mounted away from the heat zone. They are protected and see through a protective cover such as ultra violet transmitting fused quartz glass. In some situations cooling air should be blown across the UV cell face to remove dust and protect from excessive back-heat.

4.28. Notes on Firing High Temperature Furnaces

Nozzle mixing and premix furnace burners can be oversized to reduce the ambient noise level providing control is not compromised. Ideally furnaces should be commissioned using combustion analyzers, however in the absence of these tools, a slightly reducing gas/air mix with the air rate kept at a minimum will achieve the temperature required. The open type premix burners can have the primary air reduced further as a percentage is entrained as secondary air. The ideal open flame has slight yellow tips; sealed burners should be blue without hard, defined cones. Furnace pressures should be slightly positive to ensure the flue is not drawing excessively and there is no possibility of wasteful ingress of secondary air. This can be tested in a basic fashion by using a hollow tube fitted through a gap in the door. Seal the rest of the door opening completely with ceramic fibre and fit a balloon on the other end of the tube. A slight inflation of the balloon indicates slight positive pressure.

Furnace pressure is controlled by the flue and flue exit size assuming the gas and air flows are at the most efficient settings. If burners have been undersized, more capacity is possible by increasing the flue height and the draw to allow greater furnace gas inputs. Furnaces take longer to reach operating temperatures during the initial heat up phase. Allow this time to slowly heat castables or furnace parts that may not have been fired past critical temperatures previously. Check the heat up rate recommended by the manufacturer to be safe and use a pilot or small torch as a low rate if necessary.

Refer to specific instructions on setting up and adjusting burners. Nozzle mixing burners usually require special proportionators in the gas line for accurate high to low control of the flame although some furnaces require models that are able to modulate the gas only for better turn down and even temperatures. Premix burners have a gas adjusting valve under the aluminium cap located on the mixing chamber. Loosen the locknut and turn the brass screw anti-clockwise for more gas. If a zero type regulator is used on the gas line for single (air) valve control of the burner output, the adjustment of the gas valve will alter the air/gas ratio only. i.e. if more gas is supplied there will be a gas rich flame over the entire range. If manual control is preferred, this screw may be set for maximum gas only and the manual air and gas valves adjusted to alter the flow.

It is possible to reduce the noise on an open type premix burner by sealing the burner in the open port (cutting the secondary air) with ceramic fibre. The trade off will be a shortened life for the cast iron tip as the cooling secondary air is not available. The tip/s should be cleaned frequently with particular attention given to the small retention ports around the main port. They should be replaced if there is excessive corrosion or the outer casting has burnt away. Of course, a better option is to use the MP ceramic tip

5. Boiler Control Systems

5.1. Fluid Control System

Figure 5.1 shows the essential features of a system for the control of a variable such as the level of a liquid enters it. The output from liquid level sensor, after signal conditioning, is transmitted to the current to pressure converter as a current of 4 to 20 mA. It is then converted into a gauge pressure of 20 to 100 kPa which then actuaries a pneumatic control valve and so controls the rate at which liquid is allowed to flow into the container.

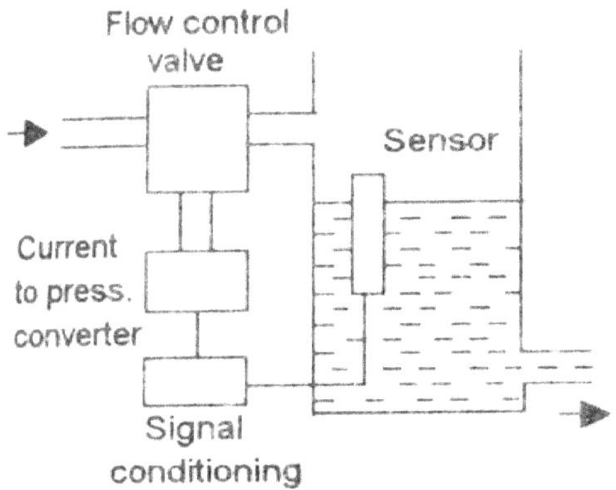

Fig. 5.1: Fluid Control System

5.2. Combustion Control

5.2.1. Combustion Efficiency

In a pulverized-fuel-fired furnace, the fuel is introduced through suitable burners in a finely subdivided form, and combines readily with the oxygen in the air to produce heat during the transit period from burner to furnace outlet. For complete combustion sufficient oxygen is required to burn all the carbon and sulphur and in addition, to convert the hydrogen to water and carbon monoxide to the dioxide form, thus ensuring that all the available heat has been extracted from carbon, sulphur and hydrogen. The theoretical quantity of oxygen can be calculated from the molecular weight of the fuel constituents, which represents the stoichiometric condition of combustion in that fuel.

However, it is not sufficient to provide the theoretical quantity of oxygen, as it is found in practice that an intimate mixture with the coal particles is not obtained. Some of the carbon will not be burnt thus occasioning a loss due to carbon-in-ash and some may only be burnt to carbon monoxide and not carbon dioxide, thus giving a loss due to incomplete combustion

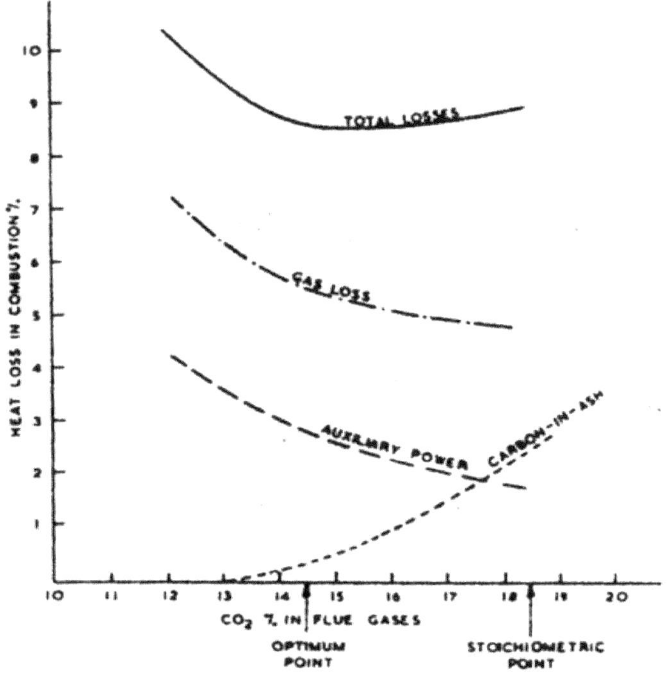

Fig. 5.2: Heat Loss Variation with Carbon Dioxide

As 1lb of carbon monoxide generates about 10,240 Btu when burnt to carbon dioxide, very small percentages of carbon monoxide in the flue gases can cause considerable loss of efficiency. In order to avoid this carbon loss it is necessary to introduce more oxygen, and therefore more air, into the furnace than is theoretically required, this being referred to as excess air; but this must be done only to the extent needed to ensure that all the carbon is burnt and that there is no carbon monoxide left in the residual flue gases. If too much air were provided it would take up heat from the furnace, which represents a loss in efficiency. These arguments are illustrated in Figure 5.2 in which it will be seen that there is a practical best position to give minimum loss of heat, known as the point for optimum combustion efficiency for a steam generator. It is this relationship of fuel-air and steam-air that forms such an important aspect of combustion control, the optimum point in the diagram being at about 14.5% carbon dioxide.

5.2.2. Excess Air

The provision of the right amount of excess air is a predominant factor in the setting of automatic controllers for optimum combustion efficiency. Carefully conducted tests on boiler plant at different loads are necessary to determine the amount of excess air required, and this can be represented by the percentage carbon dioxide in the flue gases, which is the basis for the relative factors involved in Figure 5.3. The carbon dioxide percentage will change (for the same amount of excess air) if the percentage of carbon and hydrogen in the fuel changes during the tests, whereas the oxygen content is solely dependent on the air supply. Modern control equipment, therefore, includes the measurement of oxygen in the flue gases, which can be done with better sensitivity than with carbon dioxide instruments.

It can be shown that 14.5 % carbon dioxide corresponds to about 30 % excess air, as may be observed by reference to Figure 2.4 and that this corresponds to about 5% oxygen. If carbon dioxide changes from 14.5 to 15.0 % it follows that oxygen would change from 5.0 to 4.5 %, provided that there were no carbon monoxide or other constituents present.

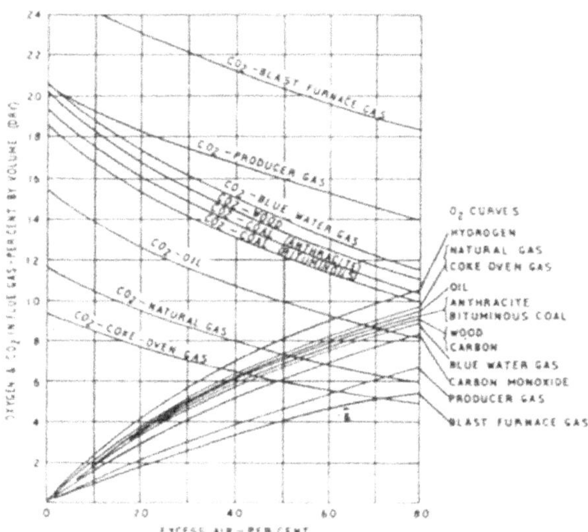

Fig. 5.3: Relationship between Oxygen, Carbon Dioxide and Excess Air

Therefore, there is much greater proportional shift in percentage gas in case of oxygen and this is used as a trimming signal in automatic combustion control. For the control of large modern boilers for 500 MW units there is a tendency to use only the oxygen signal, and this is, of course normal practice for oil-fired boilers.

Table 5.1: Relationship between Heat Release and Air Supply

Type of fuel	Fuels analysis						Theoretical air (lb of air per lb fuel)	Theoretical CO2 (%)	Heat release (Btu/lb air)
	Carbon (%)	Hydrogen (%)	Oxygen (%)	Nitrogen (%)	Sulphur (%)	Calorific value (Btu/lb)			
Coal A	76.0	5.3	16.2	1.7	0.8	13.110	9.9	18.9	1310
Coal B	94.0	3.05	1.45	1.0	0.5	15,400	11.8	19.5	1300
Oil C	85.0	12.5	2.5	-	-	18,500	14.2	15.4	1300
Oil D	85.3	10.7	4.0	-	-	18,200	13.5	15.9	1370

With the range of fuels used in large boilers, it is shown in Table 5.1 that 1 lb of air is theoretically required to release approximately 1300 Btu of heat. The actual figure varies slightly with the calorific values of the different fuels, but there is a close relationship between airs supplied and heat release. For this reason there is also a close relationship between steam flow and air flow.

5.2.3. *Firing Requirements Due to Load Change*

When considering changes of steam pressure it must be understood that the steam pressure does not change immediately on load change or in proportion to load change. The change in pressure is a resultant of load change, firing rate, and heat storage. The heat storage is contained in the water, tubes, super heaters, air heaters, economizers-in fact all the boiler parts.

From the above-mentioned facts it follows that a pressure drop cannot be restored without over-firing for the load being carried. This is in order to replace the withdrawn stored energy which was given up whilst the pressure was dropping. The reverse occurs on a rise in pressure, when the boiler must be under fired so that the excess energy is used for steam production. From all these points it can be easily seen that the time constant of a boiler is large when compared with the other possibly short time constants of a control system. This must be taken into consideration when "setting up" an automatic boiler controls system because the boiler and controlled auxiliaries are part of the control system. Their characteristics affect the response rate of the complete control system, and determine the type of control system which is needed for maintaining the stability of the boiler whilst at the same time fulfilling steam requirements.

5.2.4. Steam Flow-Air Flow Control

If a boiler is operating at its rated temperature and pressure, steam flow will be proportional to the heat output. As a result of boiler tests the losses are known and can be added as equivalent heat units to the heat output so as to give a figure for the heat input, and there is a definite relationship between steam flow, heat input and air required. Therefore, this relationship is used to control the amount of air needed for correct combustion conditions for any particular steaming rate.

When a change in load occurs, there will be a change in boiler pressure which has to be brought back to rated value. For reasons already explained in Section 2.2.3 the fuel required is in excess of the steam demand for an increased load, and vice versa for a decreased load, in order to restore the pressure as quickly as possible. The control system used for this purpose is shown in Figure 5.4 in which the role of the master pressure controller can be appreciated, as it is combined with the steam flow transmitter to alter the air flow controller.

The output from the air flow controller is applied to a metering controller which sends a signal to the fan damper positioning control or a fan motor speed regulator. The air flow to the combustion chamber is metered and a feedback signal applied to the metering controller to ensure that it gives the right adjusted air quantity.

Associated with this operation is the separate control action taken by the fuel supply which is also directly affected by the steam pressure signal as explained in the next section.

5.2.5. Fuel Flow-Air Flow Control

If the load falls suddenly there is a temporary rise in steam pressure consequent on the reduction in steam flow, and there is, therefore, a signal from the air flow controller to reduce air supply. Because of the fixed relationship between fuel flow and air flow, there is a resultant reduction in fuel input and the combined effect of these is to reduce heat release in the furnace, and thereby to reduce steam generation. This would affect the steam temperature, but the design of the boiler and super heater, together with superheat control system is such that steam temperature remains substantially constant between 70% and 100% CMR for large boilers. With oil-firing the calorific value is practically constant, and use can be made of the steam pressure signal to change fuel and air simultaneously in the correct proportions to give heat release in the boiler combustion chamber to match changes of steam flow. As the calorific value of coal varies appreciably, this matching of control with steam demand is not easy, and adjustment for coal quality has to be made occasionally.

It must be emphasized that the automatic control will supply the correct amount of air to a boiler, but the operator has to ensure that fuel supplied to the furnace is evenly distributed to give reasonably uniform temperatures across the combustion chamber.

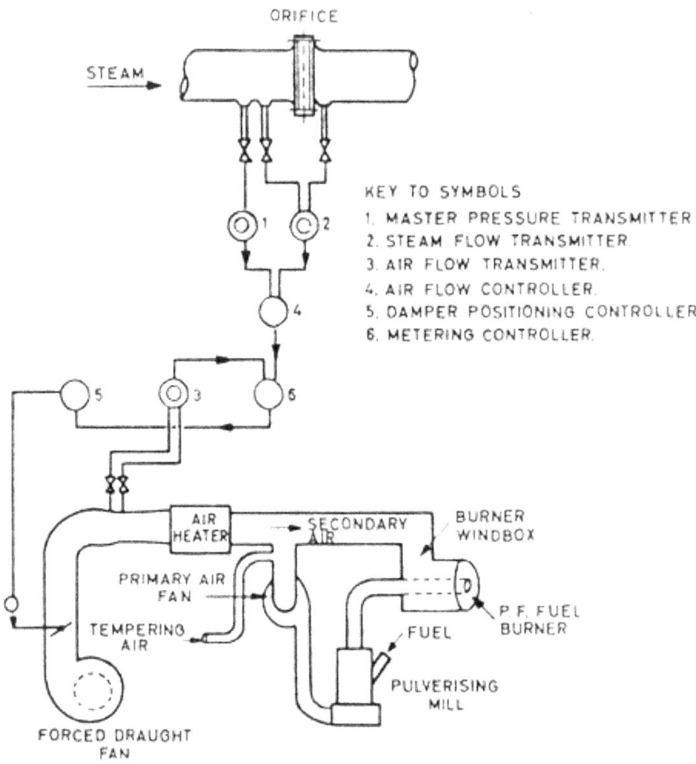

Fig. 5.4: Combustion Air Control System

5.3. Pulverized Fuel Mill Control

Coal is fed into the bunkers by conveyors, and the outlet from these bunkers is taken through a coal feeder which regulates the fuel flow to a pulverizing mill. In earlier installations this fuel feed rate was by

1. Variable position "knife" or "plough" on a fixed-feed table,
2. Variable-speed table, although
3. A two-speed table has been used.

For large boilers, about 300 MW and over, the drag or chain link feeder is used with a variable-speed drive. This permits the bunker discharge area to be sufficiently large to avoid "choking", and the link arrangement assists in obtaining an even flow of coal to the mill. The feed rate is dependent on a feeder controller working in conjunction with the air supply regulator so as to maintain fuel-air ratio.

The desired fuel-air ratio has to be determined for each mill separately at different steam 'outputs in order to set up the mill control loop and provide a characteristic line relating coal through-put to air now. The function of the mill control equipment is to maintain this characteristic automatically for optimum combustion conditions.

Four types of mill are being used by the C.E.G.B., namely:

    a.   Vertical spindle, ball or roller type, mill (pressure).
    b.   Vertical spindle, ball or roller type, mill (suction).
    c.   Horizontal type tube ball mill (suction).
    d.   Horizontal type tube ball mill (pressure).

It should be mentioned that the vertical spindle type mills are being superseded by tube ball mills for the larger boilers in conventional stations, and that these are to work under pressure of air provided by a primary fan.

5.3.1.   *Control Loop for Vertical Spindle-type Mill (Pressure)*

Figure 5.5 illustrates a mill control loop on a positive pressure type of mill.

The measurements used in the control loop are those of:

    1.   Primary air flow-a differential pressure across a restriction.
    2.   Mill outlet-inlet differential pressure.
    3.   Mill outlet temperature.

For ease of explanation assume a drop in boiler pressure, causing the signal from the master pressure controller to increase. This signal "loads" the right-hand side of the force - balance part of the primary air controller which increases the air flow in accordance with the preset mill characteristic. This air flow is measured by the orifice and a feedback signal from the primary air duct now measurement "loads" the left-hand side of the force-balance part of the primary air controller. This controller will progressively open the primary air dampers until the feedback signal rebalances the master signal. The correct amount of air demanded by the master controller will now be flowing through the mill, picking up more fuel to restore the boiler pressure.

Coal is supplied to the pulverizing mill through a coal feed controller, the pulverized coal being a mixture of coarse and fine particles, and it is the percentage of fine particles which pass through a 200-mesh sieve that is a measure of the quality of fuel. This mixture of fuel is lifted by the air flowing through the mill chamber and is carried up through a classifier to the pipes leading to the burners. There is an air pressure drop through the mill, and its value, known as mill differential, is a measure of the quantity of fuel in the mill.

In order to keep the fuel-air ratio of the mill constant, the correct amount of coal must be supplied to the mill. The mill differential pressure "loads" the left-hand side of the force balance part of the feeder controller while the primary air flow measurement "loads" the right-hand side of the force-balance part of the feeder controller. The controller is set to maintain a ratio between these two measurements, so that the correct amount of pulverized fuel will be in the mill which will cause a definite mill differential for any particular primary air flow. Under these conditions the feeder controller will be balanced.

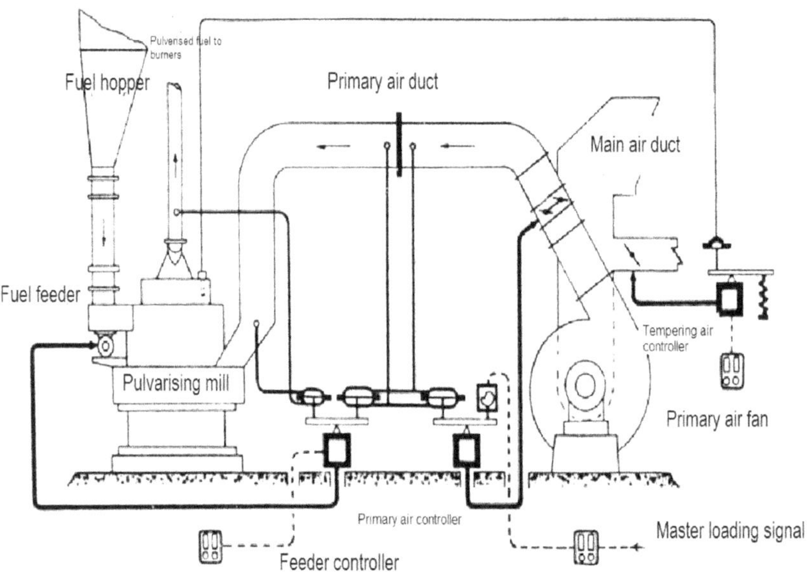

Fig. 5.5: Pressure-Type Vertical Mill Control System

If the amount of fuel in the mill is not in accordance with the fuel-air ratio required (as sensed by the mill differential-primary air flow force-balance), the controller will become unbalanced. It will then alter the feed control of coal from the bunker in such a direction as to restore the quantity of fuel in the mill to the required amount.

A temperature-sensitive element is fitted into the outlet duct of the mill, and this controls the tempering air dampers at the entry to the primary air fan in order to keep the temperature of pulverized fuel within prescribed limits to avoid mill fire. The temperature limit is set as a result of operational experience and knowledge of the characteristics of the fuel being used.

5.3.2. *Control Loop for Vertical Spindle-type Mill (suction)*

Figure 5.6 shows a mill control loop on a suction type mill.

The measurements taken to enable this control to function are:

- Mill inlet suction.
- Inlet air flow to the mill-a differential pressure across a restriction.
- Mill outlet temperature.

In this loop the exhauster controller, which controls the motor speed, serves the same purpose as the primary air damper controller in the pressure mill loop. Again assume an increased master "loading" signal due to a drop in boiler pressure. This signal "loads" the left-hand side of the force-balance part of the exhauster controller which can be characterized as before to suit the mill. The feedback signal from the mill inlet duct flow measurement which "loads" the right-hand side of the force-balance part of the exhauster controller will be inadequate, and the controller will be unbalanced, thus sending a signal to increase the speed of the exhauster until the air flow feedback signal balances the master signal. More fuel will thus be taken to the boiler by the increased air supplied by the exhauster.

The master signal is also used to adjust the feeder controller, and as the master signal increases, the speed of fuel flow to the mill increases. If the amount of pulverized fuel in the mill increases above predetermined requirements, the mill resistance to air flow will increase and flow of air will thus decrease. This will be measured by the exhauster controller which will speed up the exhausters to restore conditions. The boiler pressure will increase, causing the master signal to decrease, thus reducing the supply of coal to the mill.

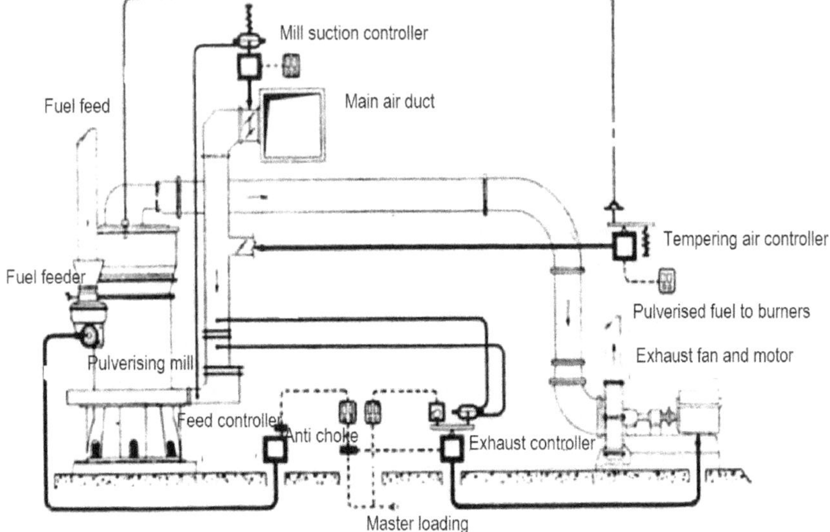

Fig. 5.6: Suction-Type Vertical Mill Control System

If, however, the amount of pulverized fuel in the mill decreases below requirements, the resistance of the mill to air flow will decrease. The flow of air will thus increase, causing the exhauster controller to reduce the exhauster speed and restore conditions. The boiler pressure will therefore drop, and the master signal will increase the supply of coal to the mill by operation of the feeder controller.

The mill inlet suction controller maintains the suction in the mill by adjustment of dampers in the supply air duct. The temperature-sensitive element fitted into the mill outlet controls the tempering air dampers as in the pressure type mill.

5.3.3. Control Loops for Horizontal-type Tube Ball Mills (suction and pressure)

These mills consist of a steel tube containing a charge of heavy steel balls which pulverize the fuel by attrition and impact action due to movement of the balls in the fuel mixture as the tube is rotated.

a) Suction type: Most of the earlier tube ball mills were suction type using an exhauster, and their control equipment was therefore similar to the suction type mill described in the previous sub-section. The arrangement is shown in Figure 5.7 which indicates that the fuel is introduced at one end of the mill together with "sweeping" air from the main and tempering air ducts at low pressure; and the pulverized coal-air mixture is withdrawn from the other end of the mill.

The pulverized fuel level is detected by means of a differential pressure provided from two tubes inserted into the mill at the outlet end. This differential is proportional to the amount of fuel in the mill, and it is therefore applied to the coal feeder controller.

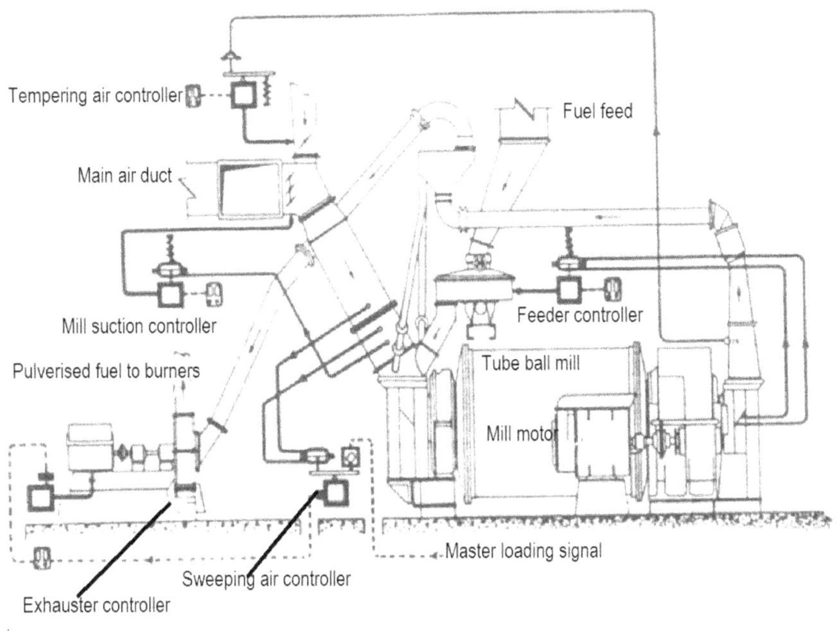

Fig. 5.7: Control System for Suction-Type Tube Ball Mill

b) Pressure type: The difficulty of providing an adequate sealing air system for tube mills has recently been overcome, and pressure-type large-capacity mills for 500 MW boilers are being installed. There are two types: one has coal through-put as in the mills already described. While the other is designed with coal entry and pulverized fuel take-off at both ends of the mill. A simplified diagram of the latter arrangement is shown in Figure 5.8 and the control is summarized as follows:

i)The output from the mill is controlled by the primary air quantity entering the mill" The p.a. quantity is controlled by an adjustable damper, the setting of which is adjusted by the signal from the p.a. damper controller which compares the p.a. required signal from the boiler master regulator with the p.a. flowing signal from the orifice plate.

ii) The raw coal feed is regulated by a coal-feeder speed controller, which responds to a signal proportional to the depth of coal in the mill given by comparing the differential pressure between points within the coal bed and on top of the coal bed.

iii) The amount of coal at temperating air admitted to the p.a. is determined by the outlet temperature of the p.f.-air mixture.

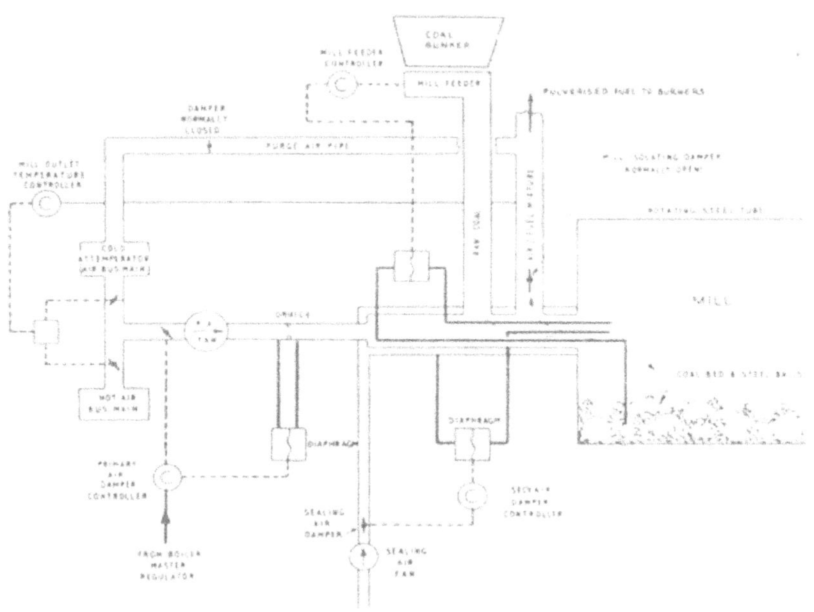

Fig. 5.8: Control System for Pressure-Type Tube Ball Mill

iv) Since the mill is under pressure there is a danger of ingress of p.f. into bearings, and escape of p.f. to atmosphere. An effective seal is provided. By using air at a pressure slightly higher than the mill pressure in a jacket around the trunnions. Control of the amount of sealing air supplied to the sealing air passages is by a damper, the setting of which is controlled by the differential pressure which exists between the mill internal and the sealing air passages.

v) The primary air flow to all mills on a boiler is summated to ensure that the correct quantity of secondary air is admitted to the burners to maintain the correct fuel-air ratio.

vi) In the event of a mill shutdown, the mill isolating damper closes and the purge air damper opens simultaneously. Air from the cold air bus main then purges the p.f. from the fuel supply pipes between the mill and the burners. After a predetermined period, the purge air damper closes.

vii) The tubes, which convey the pressures within the mill to the diaphragms, especially the one in the coal bed, are liable to become blocked with p.f., causing the control system to fail. To prevent this pressure tubes are duplicated and in addition a separate control system is provided which continually purges the various lines.

5.3.4. *Oil-burner Control*

As there are a number of different designs of oil burners, it is only possible to describe the control arrangement for one system to serve as an example. One of the essential requirements is to keep the residual oxygen in the flue gases to an absolute minimum 10 minimize sulphur-trioxide and the formation of acid condensates which can cause damage 10 air heaters and ducting where gas temperatures are below the acid dew point. Therefore close control of fuel flow-air flow ratio must be maintained at all loads preferably with less than 1 % excess oxygen.

For a wide range of load it is advisable to have a large number of burners so that each can have a low turn-down ratio. The 500 MW boilers at Fawley are each provided with thirty-two burners, arranged and brought into action in the order shown in Table 5.2 and 5.3. With a multi-burner system the burner turn-down ratio can be about 1.5-1.0 yet provide an overall turn-down of about 50-51. If the oil-pressure variation across the burner tips is proportional to the air-pressure drop across the air registers, this relationship is independent of the number of burners in action. The differential pressure between the wind box and furnace chamber governs the burner manifold pressure and, provided that the burners are kept clean, good combustion is achieved. It is usual, in large plant, to start-up with three burners, having made sure that the furnace has been purged. This would raise steam pressure and provide heat for saturation temperature. A fourth burner is added when steam is admitted to the turbine, and possibly a fifth burner to take the set up to speed. After this, the burners are lighted in pairs automatically from load demand, until all are in action giving about 70% CMR, Thereafter increase in oil pressure provides the additional heat needed up to 100% CMR.

A burner ignition controller is provided to ensure smooth increase in fuel delivery by means of a special network which calculates the oil pressure at which the additional burners are introduced. This network derives its signals from the oil manifold pressure and the number of burners already in service, using the equation.

Table 5.2: Arrangement of oil Burners at Fawley Power Station

Row A	1	2	3	4	5	6	7	8
Row B	1	2	3	4	5	6	7	8
Row C	1	2	3	4	5	6	7	8
Row D	1	2	3	4	5	6	7	8

Table 5.3: Arrangement of oil Burners at Fawley Power Station

Order of firing	Burner No.	Order of firing	Burner No.
1	A4	17	C4
2	A7	18	C5
3	A2	19	CI
4	A6	20	C8
5	A3	21	C2
6	A5	22	C7
7	AI	23	B4
8	A8	24	B5
9	B2	25	D2
10	B7	26	D7
11	C3	27	DI
12	C6	28	D8
13	B1	29	D6
14	B8	30	D3
15	B3	31	D4
16	B6	32	D5

$$P_i = \left(\frac{N+2}{1.5N}\right)^2 P_{max}$$

Where P_i = pressure at which to ignite the next pair of burners and

N = number of burners already in action (the turn-down ratio is 1.5).

5.4. Flame Supervision

Several types of flame viewing and detecting equipment have been tried, and this subject is still one which is fraught with difficulty to provide a reliable device.

5.4.1. Flame Monitoring

The growing trends towards centralized control renders it essential that some remote form of flame monitoring should be provided. Closed circuit television has been tried for this purpose, but the interpretation of the image, when not interrupted by some fault or obscurity, has been difficult, so that this method has not been adopted.

A method being tried at one of the large power stations recently constructed is shown in Figure 5.9. The operation of this equipment is based on the theory that when looking through a tube which runs down the centre of a burner, the central core of the flame is visible; and that when looking through a tube parallel with and adjacent to the innermost coal (or fuel) annulus only unburnt coal is seen as a darker space. The outer ends of these tubes are fitted with light sensitive cells, the internal resistance of which varies with the light intensity, the lowest resistance having the brightest light.

Figure 5.10 gives the values of resistances for different flame conditions, so that varying current is available for display at the control panel, and for interlocks associated with oil torches and pulverized fuel burners. The cells are connected in series so that the total resistances will progressively increase from low values at bright flame to high values caused by the pulverized coal being too far advanced. This current is used to operate an instrument calibrated in "flame position".

5.4.2. *Flame Detectors*

Photosensitive cells have been used for sensing the brightness of a flame and giving a warning as a result of flame failure.

The effect of light falling on a photosensitive cell either alters its resistance or generates an e.m.f. depending upon the type of cell. The output from a cell is usually amplified in order to obtain the power to operate alarms or trips.

Photosensitive cells are selective, in that their output per unit of incident radiation energy will depend upon the wavelength of the radiation for each type. To discriminate between flame and hot refractory several methods are used.

Amongst these are the following:

- By using a cell which is sensitive to the "blue" component of the flame, and not to reds or yellows which could come from hot refractory.
- By using a circuit which discriminates between "flickering" and steady infrared rays ("flickering" from flame and steady from refractory).
- By using a cell or unit which is only sensitive to ultra-violet rays of 0.2-0.26 μ wave band as present in all flame.

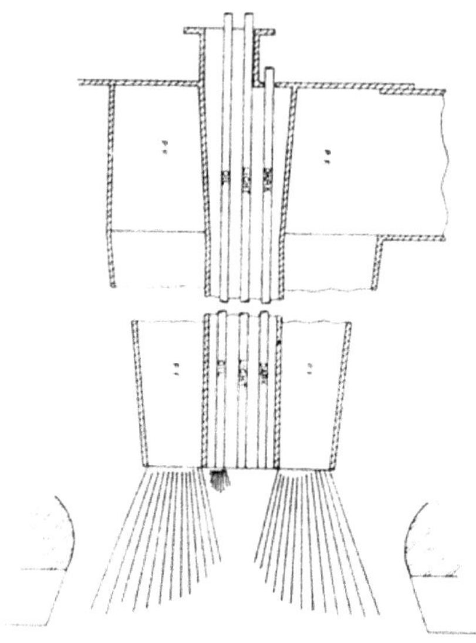

Fig. 5.9: Monitoring P.F. Flame

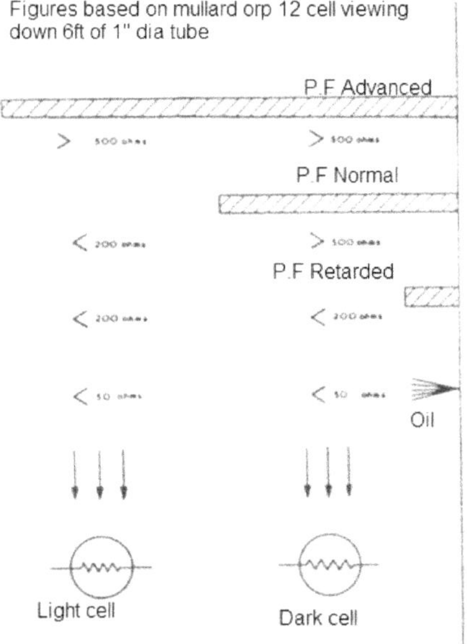

Fig. 5.10: Resistance Changes in Photoelectric Cell for "Flame Viewing"

Ultrasonic detection can be employed by passing ultrasonic sound waves down a central tube-aimed" at the flame. An outer concentric tube carries the waves "bounced" from the flame to an amplifier. This amplifier discriminates between the waves bounced from a flame, or if the flame is absent, from the furnace wall.

5.4.3.   Damper Operation and Indication

Remotely situated dampers other than those associated with the automatic control are usually operated by hydraulic or electric power units.

i) Damper Operation

The well-known Lockheed hydraulic system is illustrated in Figure 5.11 Items which are mounted on the control panel are the selector valve levers, hand-operated transmitter, and for the metering and reversing valve.

When the selector valve has connected the pipes to the appropriate damper cylinder, the lever for the metering and reversing valve is operated. These starts the motor-driven pump which supplies the operating power to move the damper, and either opens or closes the damper according to the position of the lever.

In case of pump failure, the power is obtained from the hand-operated transmitter. On large boilers an electro hydraulic system is used. The selection of a damper is made by means of a multi-contact switch on the control panel. The "OPEN" and "CLOSE" signals are also electrical and are transmitted to an electro hydraulic system housed adjacent to the boiler, On some completely electrical systems, the damper operating power is derived from a gearbox driven by a reversing motor.

ii) Damper Position Indication

When it is required to know if a damper is "OPEN" or "SHUT" the usual method is to have a switch at each end of the damper mechanism travel so that when the damper reaches one or other end of its travel a contact closes and lights a lamp on the main panel.

When intermediate positions of a damper are required to be known, an electrical poten-tiometer or similar telemetering device is coupled to the damper-operating mechanism as shown in Figure 5.12. The output from the potentiometer is fed to an indicating meter which can be calibrated to show position indication, or the indication can be of the semaphore type.

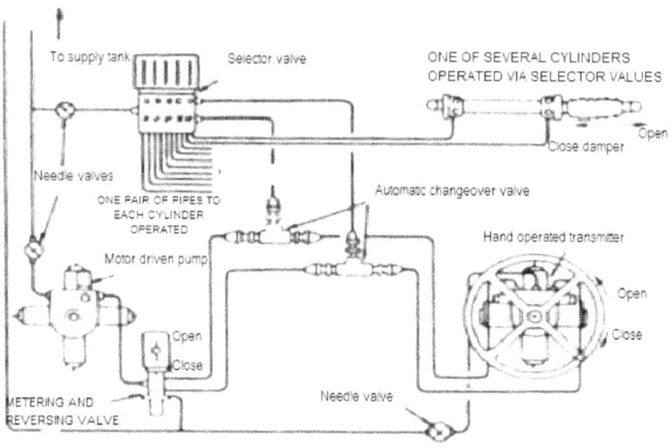

Fig. 5.11: Lockheed Hydraulic System for Damper Operation

5.4.4. Soot-Blowing

Clean furnaces and heating surfaces are conducive to high combustion efficiency, and to obtain these operating conditions it is necessary to remove ash and slag deposits at regular intervals, by steam or compressed air jets delivered from nozzles which can be inserted and retracted from the furnace. This is done by equipment known as soot-blowers which are placed at certain selected parts of the boiler walls and operated under sequence control.

Soot-blowers are located on the boiler scantlings and groups as shown in Figure 5.13 which shows that eighty-six are needed for a 500 MW boiler. Forty-six of these are on opposite walls of the combustion chamber, and these operate in pairs indicated in the diagram by a thick line half-way up the vertical scale, that is each pair initiated every 25 sec by a stepping switch and sharing about 17,000 lb/h of soot-blowing steam. Ten soot-blowers are used for the superheater, and are brought in singly at slightly differing time intervals; while the reheat sections require eighteen operated in pairs. The four economizer soot-blowers use rake-type multi-jet assemblies to get the steam distributed on several tubes, and eight operating in pairs are used for the air heater. The sequence is arranged and timed so as to cause the soot-blowers to work in the order outlined above, because the dust and deposits removed from the surfaces must be carried by the furnace gases in the general direction of the boiler exit and on to the economizer hopper and grit arrester plant. The steam supply lines to the blowers include a manually operated master valve and a remote electrically operated valve, while the soot-blowers are provided with group isolating valves.

A typical control panel is illustrated in Figure 5.14(a) which shows the locations of soot-blower groups in the superheat and reheats passes of a twin furnace boiler. Each of these locations represents a number of soot-blowers detailed on the sequential selector panel shown in Figure 5.14 (b). The selector panel is provided with a local/automatic switch and a manual push-button for individual blowers.

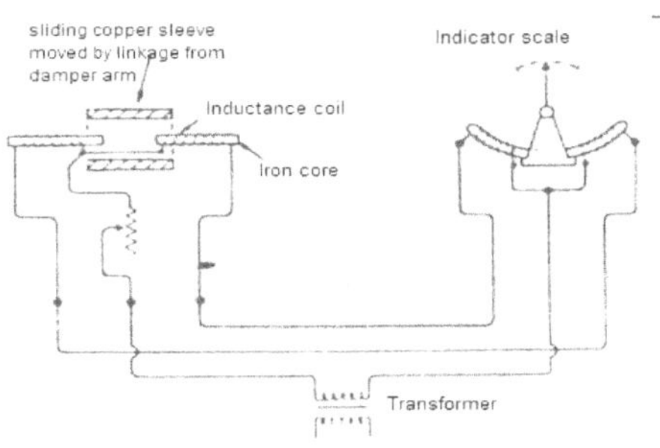

Fig. 5.12: Electrical Transmission for Damper Position Indicator

The sequence control equipment operates each blower in selected order, and an illuminated number, as shown in the centre of the control panel, indicates which blower is in operation; while the group signal light indicates which group is operating. The soot-blowers are inserted into and retracted from the furnace by electric motors which are switched from starter cabinets located near the plant, the current taken by the motor in use being shown on the ammeter at the bottom of the control panel.

The blowing sequence is started by setting the control switches for the selected groups to "automatic", and moving the main control switch to the "start" position, thus energising the group blower control units. The master valve begins to open, the isolating valves "crack" open to warm the pipelines to the blower heads, and the drain valves close when a predetermined temperature is reached. The master valve then fully opens, followed by the group valve for the first group of soot-blowers, the remaining group valves staying in the "cracked open" position. The blowers in the first group operate in numerical order, and at the end of their sequence the group steam valve closes. The next and subsequent groups proceed as above until all selected blowing is completed, after which the isolating valve closes and the drains open. Interlocks prevent the use of blowers if the steam pressure is inadequate, and stop the sequence if a blower is overloaded or operates for an excessive period. The blowing sequence can be stopped at any time by moving the control switch to the "OFF" position, and the last blower head is automatically retracted from the furnace.

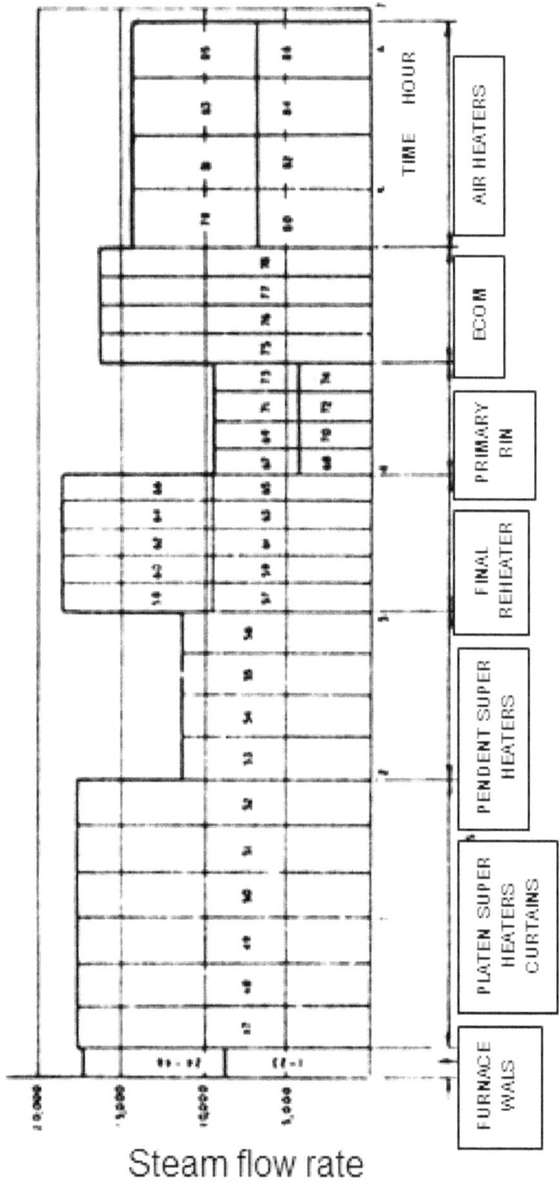

Fig. 5.13: A Sequence and Time Diagram for Soot-blower

5.5. Combustion Conditions

Whilst the automatic control systems will ensure that the correct amounts of fuel and air are supplied to a boiler, operational personnel must see that these are properly used, and view aspects of this human side of managing a boiler-turbine plant are mentioned below:

- If the pulverized fuel from the mill is too coarse, it will not be completely burned, resulting in a high percentage of carbon in the ash. The automatic control does not know this, and the signal which it receives is that there is not enough heat being supplied to the boiler, so that it automatically increases the firing rate. In fact, sufficient fuel is being supplied but cannot be used properly because of poor grinding performance of the mills.

- The correct operating level of coal in the mills must be obtained by tests and observations, and this level must be adhered to as closely as possible in order that the correct percentage of fuel in the carrying air is sent to the burners. This would be true for a particular fuel and mill operating condition, and repeat mill tests would be required for any change in the fuel or air drying capacity of the mill. It is important to appreciate this because the level of fuel affects the pressure differential of the mill which has to be taken into account when setting the mill control equipment. Too little fuel gives a transparent light flame, and too much fuel in the air gives a long smokey flame with dark "flecks", and "after burning" also occurs.

- Steam which is used for soot-blowing is not metered, and the setting of the   steam flow-air flow control is upset during this operation. This upset condition can be obviated by the use of oxygen trimming which provides a signal to compensate for the extra fuel needed for soot-blowing steam (see Section 2.6.2).

- Attention to the even distribution of air by the adjustment of dampers is necessary so that the best use is made of the combustion air which is supplied to the burners and furnace to ensure balanced and stable flame propagation.

5.6. Automatic Control of a Boiler

5.6.1. Basic Scheme

When boilers and turbines are operated as a unit, the master signal is often taken from a load measurement such as steam flow or electrical output so as to provide an open loop anticipatory control to adjust the firing system. This has an effect on the boiler pressure, and a closed-loop signal is therefore provided as a correction for steam pressure.

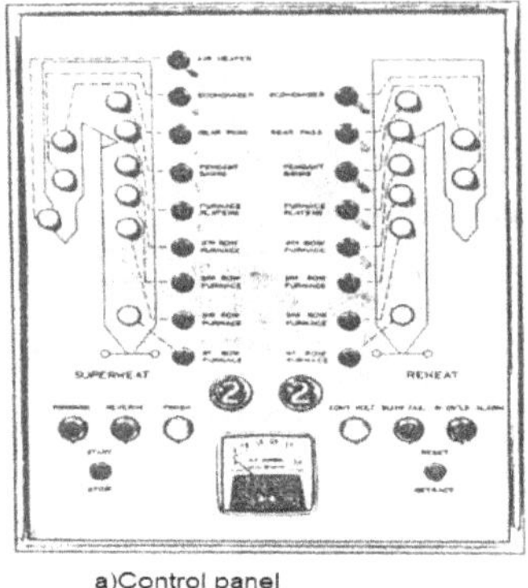

a)Control panel

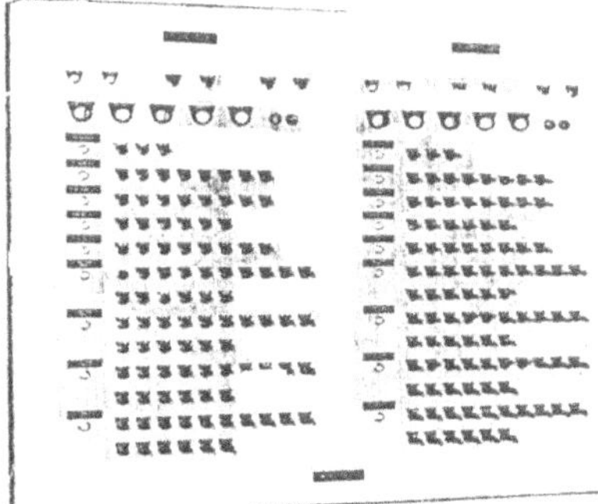

b)Unit sequential selector panel

Fig. 5.14: (a) and 5.14(b) Soot-Blower Sequence Control Panels

This compound loop consisting of the open loop for load and the closed loop for pressure correction is used for controlling the boiler. The combustion control system will respond to the reduction in steam pressure when load increases, and will subsequently restore, the thermal energy used to obtain the initial load response. From a boiler-turbine response test to a step change in speeder gear it can be shown that for a given size of generating unit operating at a load of 80 % CMR about 70% of the demanded load change can be met immediately from thermal reserve and sustained even if there is a delay in response of the combustion control system of up to 2 min.

5.6.2. *Combustion Control Scheme*

The combustion control system of a 350 MW unit is the master signal from boiler pressure is applied to mill output dampers and also as an anticipatory signal to the forced-draught and induced-draught fan controllers. The master signal is derived from the input master regulator C_1.

The temperature of the pulverizing mill output is measured and this value is compared with the set-point value in controller C_2, and the tempering air dampers adjusted to correspond, while the seal air pressure to the pressure mill is controlled by C_3 using seal air differential and set point as controlling measurements. In order to control the primary air duct pressure, controller C_4 adjusts the speeds of the primary air fans, while the primary air flow measurement signal is compared with the master signal in controller C_5 which operates the mill output control dampers, thus regulating the quantity of fuel required for meeting the boiler steam demand.

The amount of coal in the mill is controlled by C_6, which compares a primary air flow measurement with the differential pressure across the mill. A definite relationship between these measurements is established on test for the correct quantity of coal in the mill, and controller C_6 maintains this relationship by operation of the coal feeder.

Steam flow is negligible during pressure raising prior to loading, and also during low loads, and steam flow-air flow cannot be used for controlling the combustion. Therefore, wind-box pressure is used as the signal for controlling the air flow, and it is this signal which is compared with the set point in controller C_7. The output from C_7 controls the forced-draught fans. At a predetermined load or steam flow a selector switch R_1 changes the air flow control from C_7 to C_8, the latter being provided with two signals, one from forced draught air to the combustion chamber, and the other from the steam flow-air flow controller so as to trim the air supply in accordance with steam demand.

When the turbine has been run up to speed and synchronized, a "block" load is applied to the generator, and the set gradually loaded. As 'explained in the section dealing with combustion, the steam flow-air flow signal is used for controlling the boiler on load, while the air flow signal introduces the heat control by means of the fuel flow-air flow controls associated with the mills and/or firing equipment.

During load control, the quantity of air supplied from the forced-draught fans is measured and this air flow signal is applied to the controller C_8.

It was mentioned earlier that the main load control is conditioned by measurement of steam flow, and a signal from this measuring device is applied to controller C_8 through an oxygen trimmer. In order to compensate for transient conditions and to ensure the best possible combustion at all times, the flue gas is continuously monitored for residual oxygen, and if there is any deviation from the desired value of oxygen in flue gas, the analyzer will feed a "trimming" signal into the steam flow signal chain. Furnace pressure is maintained by a controller C_9, which controls the induced draught in relation to forced draught to give a negative pressure or small suction in the combustion chamber. In order that the forced-draught and induced-draught fans respond quickly, an anticipatory signal is applied from the boiler pressure through the master regulator C_1.

5.7. Setting up a Control System

Frequency response analysis by injecting sinusoidal disturbances requires more equipment and takes longer than plotting recovery curves resulting from the injection of step disturbances, but the former method gives data from which the frequency response characteristics of individual controllers or control loops can be obtained, and the results used to predict the behavior of a complete control system. To obtain frequency response data, a sinusoidal input is applied to the control element or loop. Two main reasons for using an input of this nature are:

a.   It facilitates the comparison of mathematical deductions and results.

b.   Most restoring forces in a control system have a simple harmonic motion.

Figure 5.15 illustrates the method of obtaining frequency response data in a fuel control loop. The loop is opened at an arbitrary point and a suitable sine-wave generator (pneumatic, hydraulic or electronic) provides the input signal to the loop, in place of the master or control signal. The input signal to the control loop is recorded by one pen of a high-speed two-pen recorder. The second pen records the return signal from the loop. Suitable amplitude of input signal is chosen bearing in mind the speed of movement of the components. This is an important point because a system may have a large time lag without any attenuation and yet he stable at high frequencies if the signal amplitude is small. A number of tests are conducted at different frequencies, the results being graphed by the recorder. The Figure 5.16 shows the response curve of a system having no lag and a gain of unity, and it can be seen that the output is a complete reversal of the input. This is a phase reversal which is due to controller action and must not be confused with phase lag. Figure 5.17 shows the response of a system with a lag of 90° and a gain of 0.6.

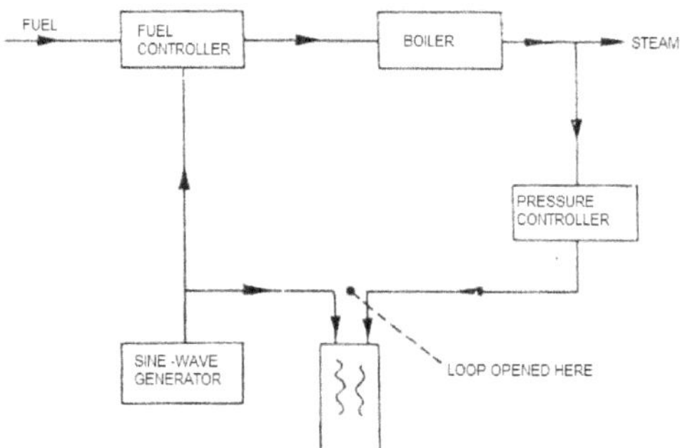

Fig. 5.15: A Block Diagram of Control Loop under Test

When a number of tests have been conducted the results are extracted from the charts and a Bode diagram or a Nyquist diagram can be constructed, from which conclusions can be drawn regarding the stability and behavior of the control system and plant under various conditions. If the responses of the loop are unsatisfactory, corrective adjustments can be made and the tests repeated as necessary. It should be pointed out that the frequency response of the test equipment must be much higher than the test frequencies used for the analysis. The straightforward "opened loop" is of great practical help in setting up a control system or investigating causes of its stability during "trouble shooting" or commissioning of control equipment. When more complete data is required. complex tests with highly instrumented test rigs, data loggers and computers have to be used.

Under practical operating conditions, the accepted approach to tuning the plant control system has been to adopt the time-tried techniques of increasing the gain of the proportional control until the system just oscillates and then reducing gain and setting integral time in relation to the period of oscillation. Unfortunately, this technique usually relates to one definite load, and unless detuned takes little account of interaction between the major loops which result from subsequent load changes. In order to try and save as much time as possible on site, a design study working party has made a special examination of the control schemes for the Fawley power station, and a detailed analogue design simulation has been made using the C.E.G.B. computer. From the simulated computer plant runs at 30, 50, 70 and 100% CMR it has been possible to provide pre-commissioning guide to correct control system tuning, and this may form a pattern for the setting-up of control equipment in future stations. A useful rule of thumb method for setting a "three-term" controller is to switch out all integral and derivative action and increase the proportional gain until a continuous constant amplitude oscillation can be simulated. Then reduce the proportional gain to half its setting, set the integral action time equal to the period or the oscillation and the derivative action time to one-eighth of the integral action time. A decay ratio of one-quarter should then be set in by marginal adjustment of all three terms.

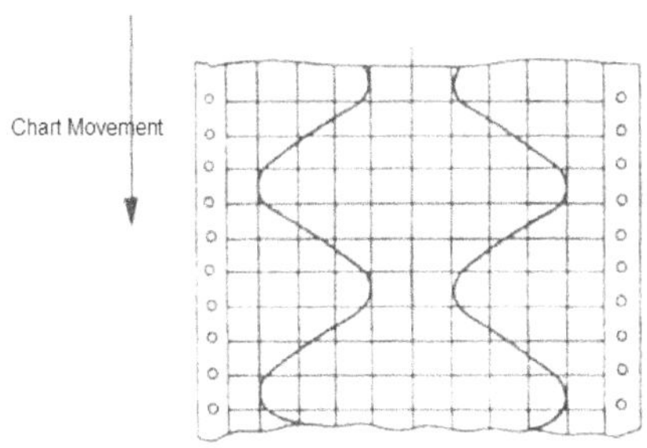

Fig. 5.16: System with 0.6 Gains and 90° Lag

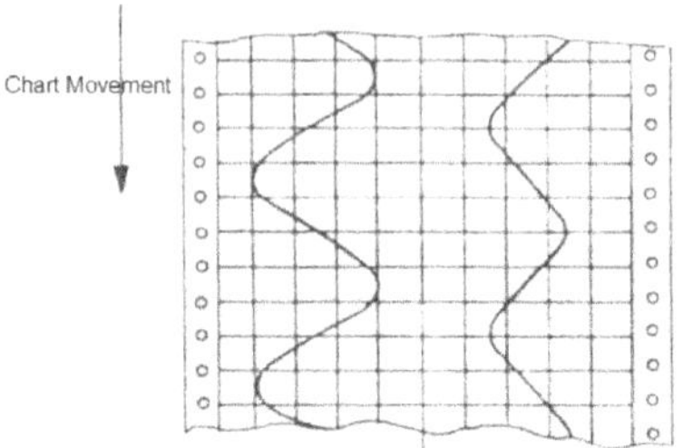

Fig. 5.17: System with 0.6 Gains and 90° Lag

5.8. Steam Temperature Control

5.8.1. General Requirements

In any modern thermal power station, it is of the greatest importance to keep a very close control of the steam temperature at the boiler outlet, since excessive temperatures have a serious effect on the metal endurance of all parts of a boiler or turbine which are subjected to it. On modern large steam turbines the temperature must be held close to design values and excessive temperature gradients avoided, for the following reasons:

a.   Small clearances between stationary and moving parts require limits for rotor movement relative to the turbine fixed blading.

b.   Steels working in the region of 565°C have lower safety margins due to "creep" stresses at higher temperatures.

c.   Close matching of steam-metal temperatures is necessary, especially during start-up and shut-down.

Superheat and reheat, control has to be considered for a load range from no load to full load. There are several methods for controlling the steam temperature, and their use has to be decided with due regard to boiler design and the requirements of turbine starting. A combination of methods may be necessary; temperature boosting may be required at some loads, whilst desuperheating may be needed at other loads, and sometimes boosting of reheat accompanied by desuperheating of main steam at low loads.

5.8.2. Heat-exchanger-type Desuperheater

For the comparatively small turbines this type of desuperheater is used for superheat control when it is known as a non-contact type desuperheater. It is usually placed between the primary and secondary super heater sections, the steam passing through a nest of inverted V-tubes and the water which is taken from the boiler-feed circuit flows around these in the body of the exchanger in which the tube nest is accommodated. The control is affected by the intermittent operation of bypass valves on the steam side, but as this has caused corrosion troubles this method of control is not now used.

Heat exchanger type desuperheaters are, however, used for reheat control for some large turbines now being installed, but for these applications the water passes through the tube nest located in a steel barrel through which the steam is passed. The control is by means of steam bypass valves which are continuously adjusted to obtain correct temperature.

5.8.3. *Spray-type Desuperheater*

In order to obviate the stresses on large joints as used on the non-contact desuperheater, water from the feed supply to the boiler can be sprayed into the steam, as shown in Figure 5.18. The spray desuperheater is fitted with an inner sleeve to prevent water coming into contact with the tube wall.

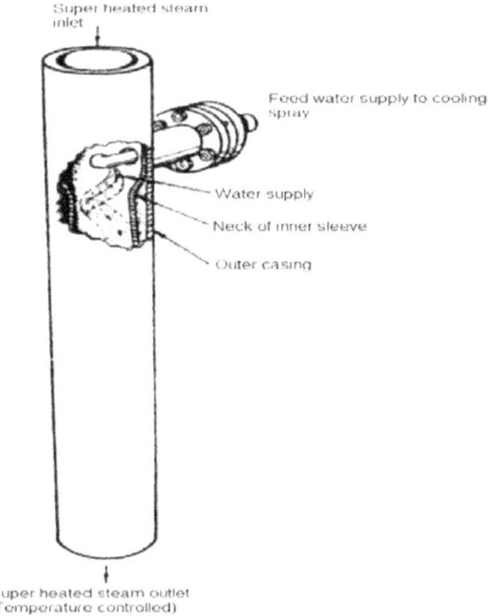

Fig. 5.18: Spray Type Desuperheater

A temperature-sensitive element in the main super heater outlet sends a signal to the spray water valve controller. The controller adjusts the spray water admission valve in response to the signal. A flow measuring device in the spray water line, sends a feedback signal of flow to the controller. Thus the correct flow of water is admitted as demanded by the controller.

5.8.4. *Tilting Burners or Burner Pattern*

Two methods of controlling the furnace heat absorbed by the super heater or reheater are tilting the burner nozzles or changing the pattern of firing by burner selection. Two different types of pulverized fuel burners are shown in Figure 5.19(a). A hand wheel for nozzle tilting is shown in the right-hand photograph while automatic mechanism for remote control of burner nozzle adjustment is clearly shown in the left-hand illustration. The range of control of heat release is given in the diagram of figure 5.19(b)

Burner selection can be used to influence the heat transferred to the steam. If firing is organised to use mills which supply the top rows of burners, additional superheat is obtained. The pattern of firing is selected to give best combustion and heat transfer of the steam generating and superheating surfaces.

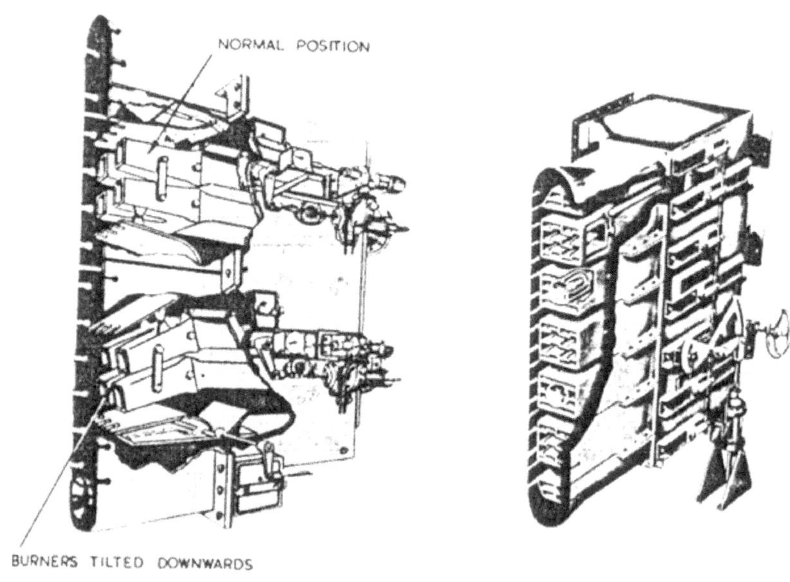

Fig. 5.19: (a) Tilting Corner Burners

5.8.5. *Super Heater Gas Bypass Control*

This controls the heat absorbed by the heating surfaces in the convection zone, by means of bypassing some of the hot flue gases away from the super heater, as illustrated in Figure 5.20. The bypassed gases may pass through a part of the economizer in order to recover the heat they contain.

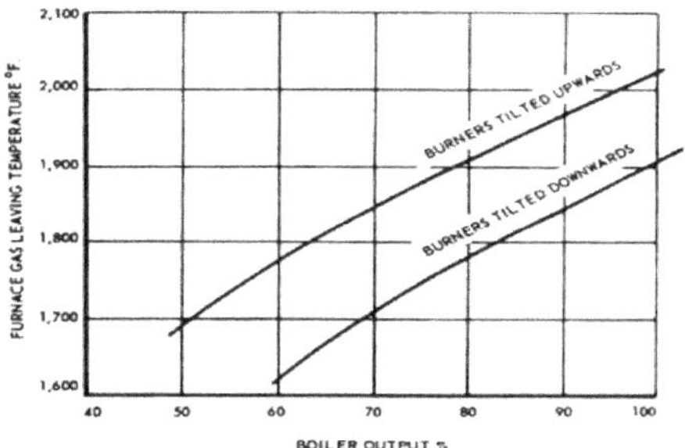

Fig. 5.19: (b) Variation in Furnace–Leaving Temperature Resulting from Burner Tilt

The temperature controller in the main super heater outlet senses an error in steam temperature, and re-positions the bypass damper to restore the required steam temperature.

5.8.6. *Gas Recirculation for Boosting Super Heater*

Gas recirculation is used to increase the mass flow of hot gases through the super heater by withdrawing gas from the gas outlet of the economizer by means of fans and discharging it into a manifold around the bottom of the furnace chamber, thus tending to cool the walls and increase heat absorption by super heaters. The hot gases pass through the furnace; mix with the spent gases above the flame zone, thus increasing the mass flow of hot gases through the super heater without any increase in combustion air needed. The automatic control selects the number of fans required to boost the temperature and it is usual to have this method of steam temperature control during a "start-up" of a boiler turbine unit.

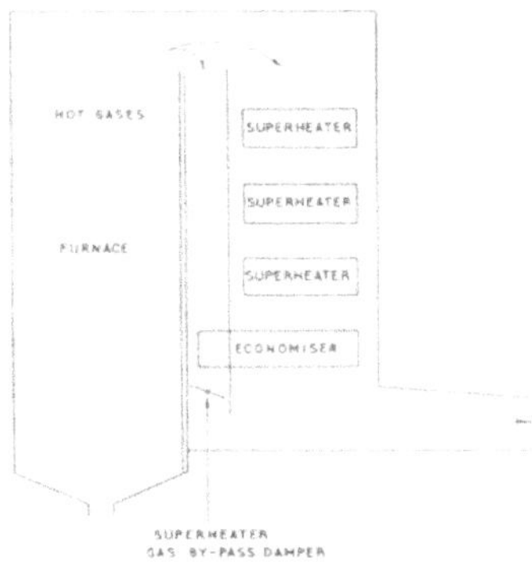

Fig. 5.20: Super Heater Gas by Pass for Steam Temperature Control

5.9. Steam Temperature Control Systems

5.9.1. Superheat Control

Spray type desuperheaters are used in a cascade system of control as shown in Fig. 5.21. The final steam temperature is measured by a potentiometer at 1 and a signal sent to the controller 3, where it is compared with the set point. The output of 3 is used as the set point of controller 4. This set point must, therefore, vary if the super heater outlet temperature varies.

The primary super heater outlet temperature is measured by a potentiometer at 2, and a signal sent to controller 4, where it is compared with set point from 3. The output signal from 4 goes through relay 6 to an auto-manual changeover point, hence to the spray water admission valve.

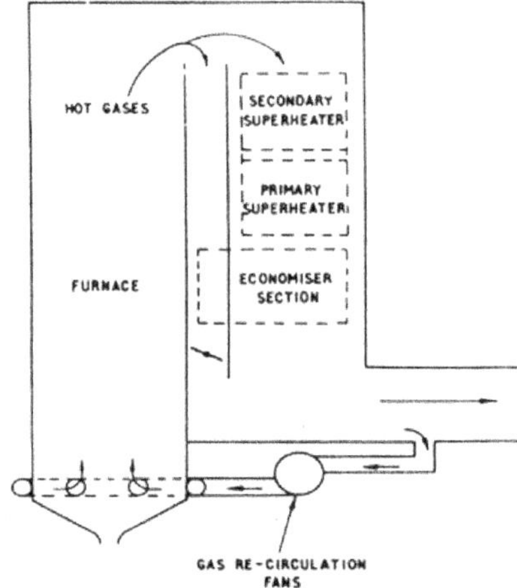

Fig. 5.21: Steam Temperature Control with Super Heater Gas Bypass and Gas Recirculation

The effect of the variable set point from 3 can now be seen. It is used so that the temperature at 2 will be such that the temperature at I is correct, because if the temperature at I is in error, 3 will change the set point.

Relay 6 is used so that immediately any change of flow is measured at F, an anticipatory signal is given to the spray water valve. Closer control is thus possible, as a "boost" is given, followed by the control loop explained above.

To understand the operation of this scheme, assume an increase of super heater outlet temperature above the desired value which is determined by the set point at controller 3.

The increase will be sensed at 1 and fed into controller 3. The difference between steam temperature and set point will cause controller 3 to modify its signal to controller 4. This is the new set point of controller 4. calling for a reduction in temperature at 2. The measurement at 2 compared with the new set point of 4 will cause the output of 4 to increase the spray water, thus bringing the temperature at 2 to that called for by controller 4.

If for any reason the temperature at 2 changes, with the temperature at 1 correct, controller 4 will signal to the spray valve to make correction before the change is felt at 1. In this way a quick response or anticipatory action is obtained from the steam temperature control system.

5.9.2. *Superheat and Reheat Control*

An inspection of Figure 5.22 will show that the superheat section is controlled as before by spray water injection between the primary and secondary super heater sections, while the reheat has dual control consisting of a gas bypass for its primary section and spray water injection between the primary and final reheat sections.

With the gas bypass dampers closed, maximum heat transfer to the reheat primary section takes place to "boost" the reheat temperature particularly at low load. The spray water injection is used to balance this heating effect by tending to cool the steam, that is to "buck" the reheat temperature (the words "boost'" and "buck" having opposite meanings).

It can be seen from Figure 5.23 that the superheat and reheat control systems are separate, and that a set point is situated on the superheat temperature regulator 2. The signal from this regulator is passed through a load change relay 3 to the spray water flow regulator 5 which obtains its flow signal from flow meter 4, and control valve 6 connected to injection nozzle 7.The reheat temperature is controlled by gas bypass dampers represented by a valve in the diagram at II, which responds to a signal from regulator 9 through load relay 10 which may provide a function representing the actual output from the boiler-turbine unit. The spray water for reheat control is adjusted by regulator 5 (on right-hand side of the diagram) so that it operates in the manner already described, a set point being introduced at the regulator station.

5.10. Draft Control

Mechanical draft fans have an important place in the engineering of power plants. Without fans the high rates of heat transfer now possible could not have been realized nor would the .thick fuel beds of underfeed stokers have been possible. Mechanical draft may be classified as forced or induced, the former having the combustion air placed under a plenum, the latter referring to gas movement produced as the result of a vacuum. The forced draft fan draws in air from the atmosphere and delivers it through air ducts either directly to the combustion equipment or to the air inlet of an air preheater. Forced draft alone is undesirable. Furnace doors or ports may not be opened without an outflow into the boiler room. Furnace gases escape through all joints and cracks in the setting. There is more "soaking up" of heat by the furnace walls. However, some package oil-burning boilers use it successfully.

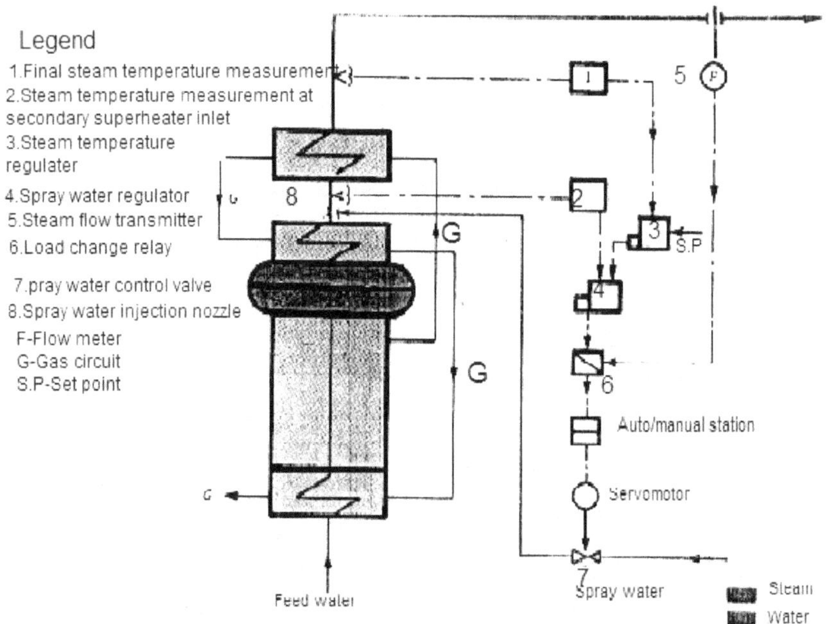

Fig. 5.22: Superheated Steam Temperature Control Cascade System

Conversely a system entirely vacuum in character will have considerable dilution of the products of combustion by infiltration of boiler room air through the setting and at any point in the external flue gas passages where the casings are not airtight. Most hand-fired boilers, also a few stokers and gas burning units are of this nature. Induced draft is created by chimneys and by fans located in the gas passage on the chimney side of the boiler and its auxiliaries. The logical arrangement is to employ both vacuum and plenum in such proportions that the furnace pressure is nearly atmospheric, and then the effect of small leaks in the setting is negligible. In a balanced draft system, the controls are usually set to maintain about 2.5 mm water vacuum over the fuel bed or in the furnace.

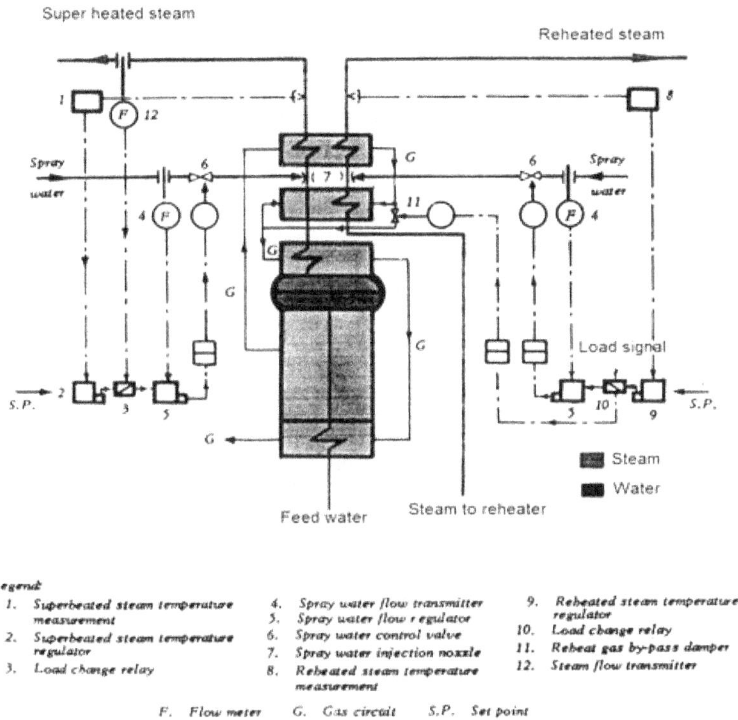

Fig. 5.23: Superheated and Reheated Steam Temperature Control

The centrifugal fan is a machine for moving large volumes of air against, a small plenum. It has a bladed wheel rotating in a stationary scroll casing. Air admitted around the hub flows into the moving blades and is thrown outwards into the scroll casing. The drive is most frequently an electric motor, either direct-connected or belted, although sometimes steam engines or turbines are used. The energy used by a fan impeller goes into bearing friction, flow work, compression, velocity head, and heating. It is commonly assumed in fan work, where the pressures are usually less than 25.4 cm water plenum or vacuum, that the fluid is incompressible. This assumption, together with considering heating as a mechanical-type loss, simplifies the work equation to

$$W = v\Delta p + \frac{V_2^2 - V_1^2}{2g} \quad \text{kg m per kg air flow} \quad \text{---- (1)}$$

in which W = Work added to the air, kg m per kg .

v = Specific volume of the air or gas, m^3 per kg

Δp = Static pressure increase, kg/m^2.

V_2, V_1 = Velocity, m per sec.

From equation 1 we see that the work, and consequently the power, imparted to the air consists of two components, Viz., static pressure and velocity, an ordinary manometer reading on the inlet and outlet of a fan gives the static pressures relative to the atmosphere. Their difference is the working static pressure of the fan. Similarly an efficient pitot tube turned into the air steam furnishes an indication of the sum of static pressure and velocity head on the attached manometer. This is called total pressure, or dynamic pressure.

If the fan sucks air directly from the surrounding atmosphere this total pressure is the dynamic draft produced. If there is an inlet duct, the before and-after pressures must be read and subtracted or taken from a manometer that is differentially connected.

Let Δp_a = Static draft, cm water

ΔP_t = Total draft, cm water

Q = Gas flow, cfm.

Then from equation 1 the following can be derived:

$$\text{Total air hp} = Q\Delta p_t /450 \qquad \text{---- (2)}$$
$$\text{Static air hp} = Q\Delta p_s/450 \qquad \text{---- (3)}$$

The fan efficiencies are expressed as follows:

$$\text{Total fan mechanical efficiency, } \eta_{ft} = \frac{\text{Total air hp}}{\text{Shaft hp}} \qquad \text{---- (4)}$$

$$\text{Static fan efficiency, } \eta_{fa} = \frac{\text{Static air hp}}{\text{Shaft hp}} \qquad \text{---- (5)}$$

In fan practice the term manometric efficiency is employed to describe the effectiveness of a fan in producing the draft pressure of which its design is theoretically capable.

$$\text{Manometric efficiency} = \frac{\text{Draft actually produced}}{\text{Theretical draft}} \qquad \text{---- (6)}$$

Some manufacturers base this efficiency on a radial blade standard. In effect, this is to assume a theoretical draft standard of u^2/g. On this basis, the manometric efficiency (better referred to as manometric ratio) of a good plate fan is 0.65; of a forwardly curved multivane,

1.12; of a backwardly curved multivane, 0.385. For high-speed motor or turbine' drive, low manometric ratios are desirable.

Draft fans are designated as plate (paddle wheel), multivane, or propeller type. The propeller type is seldom used, since it develops but little static pressure. The plate fan is employed to some extent, especially with engine drive, but the multivane centrifugal fan is the common type for both forced and induced draft. Multivane fans are constructed with blades which are radial, backwardly curved, or forwardly curved. The blade curvatures are most important in determining the fan characteristics.

The effect of blade curvature maybe studied with the aid of Fig. 5.24 in which u = peripheral speed of the blades, while V and V_r represent, respectively, the absolute and, relative velocities as the air leaves the blade.

Obviously V' must increase with increased rate of discharge. In the vector diagrams the vectors V' represent the higher discharge rate' and one may see that for a given increase in V'_r the vector V' is greater than V in two of the types, about the same in the other. Thus by study of these diagrams one may understand why the backwardly curved type possesses a self-limiting demand for power and a limited rate of discharge, whereas power increases rapidly with discharge for the forwardly curved type. But since V is largely diffused into plenum in the scroll case, it is apparent that lower "values of 11 will produce the same plenum in a forwardly curved fan.

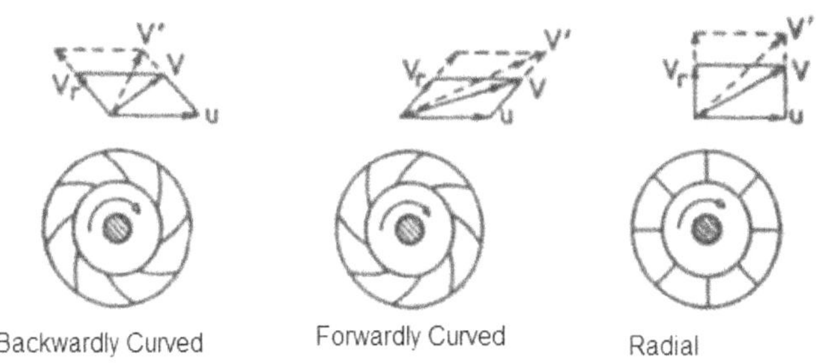

Fig. 5.24: Basic Blade Forms for Fan Wheels

Backwardly curved blade wheels are generally selected for forced draft service because the high speed is suitable for standard motor drive. The power demand is self-limiting, and the static efficiency is high. These fans may be satisfactorily operated in parallel.

Induced draft fans operate in gas of much higher temperature and may handle gases laden with dust. Forwardly curved blade wheels run at the lowest speed to develop a given pressure, hence are frequently chosen for induced draft service so that the centrifugal stresses in the wheels will be least. Low speeds, together with absence of dust-gathering tendency, minimize out-of-balance vibrations. The forward curvature reduces "the blade depth, but gives a large inlet opening for the gas. The rising horsepower and pressure characteristics of this type usually render it unsuitable for parallel operation, for it can overload its driver under abnormal working conditions. Induced draft service is exacting in requiring heavy-duty construction and is frequently met by a modified radial blade.

Basic operating conditions of forced draft service are:

1. The fan handles cool, clean air.
2. The fan location can be wherever convenient. Ducts carry air from fan to plenum chambers.
3. General use is made of backward curving, high speed multivane fans.
4. Required draft consists of air duct, preheater, and fuel bed resistance.

Induced draft service is not as simple and direct as forced draft. Its basic operating conditions are:

1. The fan handles hot gases, often from 260° – 480°C.
2. The gases often contain soot which fouls the blades or ashes and cinder which wear them.
3. Required draft consists of the sum of gas friction loss through furnace, boiler, super heater economizer, dust collectors, and air preheater. Chimney may assist.
2. Location is fixed, somewhere between boiler outlet and chimney.
3. The handling of hot gases requires more expensive construction, such as shielding or water-cooling of bearings, etc.

Vacuum costs more to create than plenum so the induced draft is used alone only with installations where little or no fuel bed resistance is encountered.

Fan manufacturers test their product under standard conditions and publish the results in tables or as performance charts. The power plant engineer's work is then the fitting of his own particular conditions with the aid of these charts or tables, making the proper corrections where his data differ from the manufacturer's test conditions. The purchase of large fans is commonly made upon the basis of performance guarantees. The generalized performance curves of Fig. 5.25 are for constant speed, since that approximates the usual condition of use.

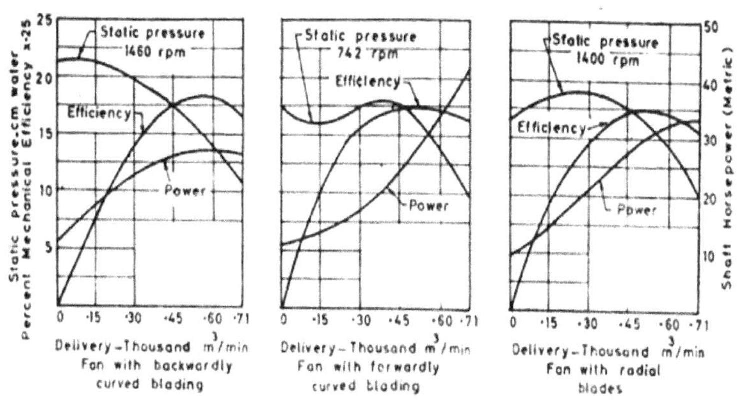

Fig. 5.25: Comparative Performance Characteristics of Centrifugal Fans

All Rated 566.34 M³/Min (20.000 Dm) At 15.24 Cm (6") Sp

But Not All Same Speed or Size

One is able to predict the performance at other shaft speeds and gas conditions, making use of such data. From fan theory, assuming efficiency to remain constant, it is possible to formulate a number of equations covering the before and after a change situation.

5.10.1. Inlet vane Control

This is most using me no in the forced draft field. The inlet vanes are located on the inlet to the fan and by adjustment can change the direction of air entering the wheel. Over limited ranges of control, say from 50 % to 100 % rated volume, there is not much throttling; rather, the inlet vanes reduce the volume of air handled by reducing the speed of the fan wheel relative to the entering air. At still smaller flows the nearly closed vanes have considerable throttling action and the performance is not much different from plain damper control.

5.10.2. Speed Control

Fans may be driven by turbines, variable-speed motors, or constant-speed motors with variable-speed couplings-hydraulic, magnetic, or mechanical. This control is, of course, far more expensive in first cost than other forms, but the resulting performance .as depicted in Fig. 5.26 quite favorable if the fan is operated much at fractional loads below 60% of rated capacity. This type of control provides more advantage on induced draft service, where inlet vane control is not very expedient.

117

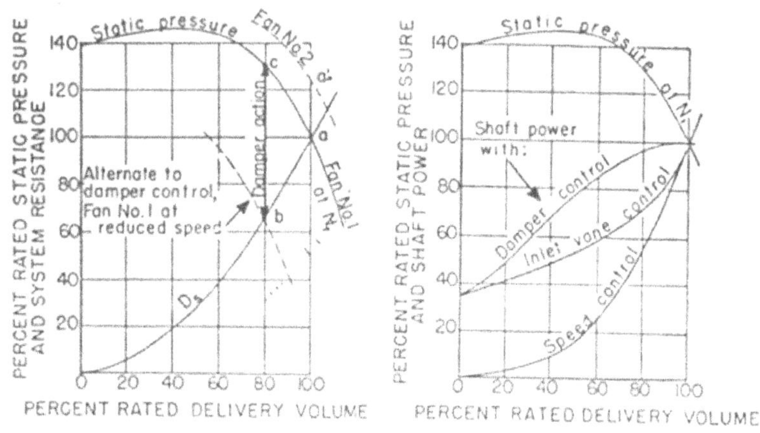

Fig. 5.26: Variable Load Performance of a Centrifugal Fan

5.10.3. *Damper Control*

This is the simplest method and least expensive in first cost; also, the least efficient on account of the irreversible, entropy-increasing, action of throttling flow. Nevertheless it is a common method of control of induced draft in power plants. The damper is usually located in the system on the boiler side of the fan. The fan outlet is an unrestricted discharge to the chimney. Since damper control merely imposes a controllable pressure drop in the system, the fan can produce point b (Fig 5.26) conditions by operating at point C with the damper increment be added to the system resistance.

5.10.4. *Construction*

Centrifugal draft fans consist primarily of a rotating bladed wheel enclosed in a spiral-shaped sheet-metal housing. The wheel may be single-entry, meaning that air enters the interior of the wheel from one side only, or double-entry. The latter form is employed for large volume fans where the necessary wheel width is so great that it is expedient to feed in the gas or air from each end of the wheel. The shaft on which the wheel is mounted turns in bearings which are supported on brackets fastened to the housing or on pedestals independently mounted on the foundation. Bracket support is the cheaper and is suitable for close-coupled motor drives. The pedestals are preferable for induced draft fans and for forced draft fans driven by turbines or speed controllers. The drive may be V-belt, but direct drive is usual, there being interposed between driver and fan a flexible coupling.

5.10.5. *Control of Gas Loop Flows*

The flow of the gas loop fluid, air or flue gas must vary if the rate of combustion varies. Because of variable demand, which is the common experience, the energy originated by combustion in heating and power plants is wanted at varying rates of production. This is true in practically all cases. It is in the Gas Loop that the energy is originated as heat; hence control for variable rate of energy production begins in this part of the plant.

Combustion being the source of heat energy, the control is primarily combustion control. A control system seeks to attain the following objectives:

1. Regulate input of heat energy to the plant equipment so that it will always be equal to the plant needs, but not in excess.
2. Maintain high efficiency of combustion at all rates.
3. Be sufficiently sensitive, so that the thermal state of the plant equipment does not fluctuate. Only the rate of energy flow should vary.

The plant operating characteristic commonly used for "mastering" the control of combustion is steam pressure. Alternately steam flow, generator amperes, or some other quantity that is affected by load could possibly be used, but steam pressure control is the common practice. When plant load increases, steam flow out of the boiler will increase, and unless the rate of combustion immediately increases the energy deficiency will quickly show up as a drooping steam pressure. Since elements like diaphragms and Bourdon tubes can be built to sense this change almost at once, they provide a convenient method of feeling out the variable load and translating it into a call for increased combustion. The necessary allowance for pressure variation over the full operating range is within 4 %, that is to say combustion control equipment can be built that will function satisfactorily from steam pressure only if a pressure-load characteristic such as Fig. 5.27 is permissible. This is substantially "constant pressure" control, but not exactly. Where the pressure must be more closely regulated the simple pressure-responsive elements must be aided by other impulses, say from flow elements. Attempts to make the pressure-load. Characteristic too flat are followed by "hunting"; hence the small pressure variation is a stabilizing feature.

In general, the control of combustion for variable load needs to adjust the fuel input proportional to the load. This must be accompanied by a change in air flow to the combustion equipment so that the most efficient air-fuel ratio will be maintained at all loads. Some secondary features usually found to be desirable are the maintenance of balanced draft, and the apportionment of load, at will, among boilers operating in parallel on the same steam main.

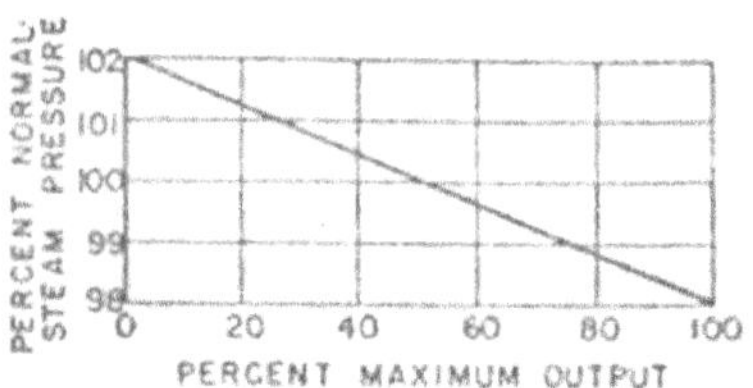

Fig. 5.27: "Constant Pressure" Control

It is possible for alert attendants, aided by instruments and gauges that inform them of conditions in the steam generator, to adjust coal -feeders, dampers, motor speeds, etc., so as to maintain steady steam pressure, optimum air-fuel ratio, and atmospheric furnace pressure while the rate of steaming varies. It is not likely, however, that this will be consistently and reliably accomplished by manual operations; also it becomes increasingly difficult as steam generators approach the "flash" type, and the load becomes more variable.

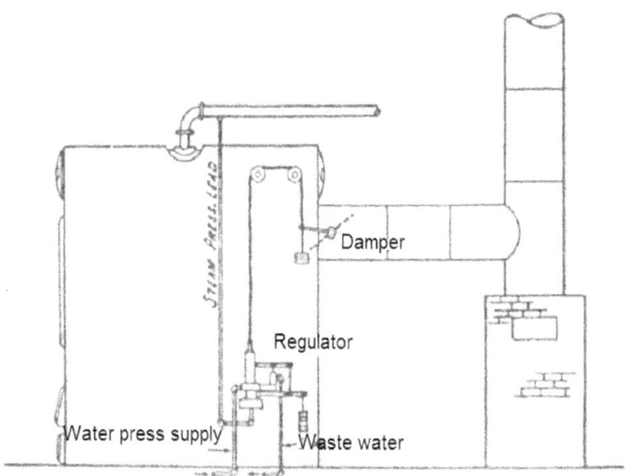

Fig. 5.28: Elementary Combustion Control by Damper Regulator

Automatic devices are built that can perform the necessary regulating actions far more perfectly than humans. These are in widespread use in heating and power plants. Varying degrees of control are possible, so that the extent to which combustion of a fuel is put under automatic control can be suited to the size of the plant and the money available for control installations.

A simple form of control wherein steam pressure automatically adjusts the draft is applicable to the smallest plants. The damper regulator shown in Fig. 5.28 substitute's mechanical control for the frequent opening and closing of dampers by competent firemen, or for the open position often maintained by poor firemen. The rate of fuel feed remains under manual control; however, this phase of combustion control can be better judged by visible indication than may gas flow.

From this simple case the degree of automatic control available progresses to the other extreme of completely automatic control capable of successfully operating several large boilers in parallel with peak combustion efficiency in spite of fluctuating load conditions. Lack of space prevents a complete description of, all the systems that have been employed as there is a multiplicity of detail and equipment in this field. However, the principle of complete control will be explained, then illustrated by a few chosen examples.

We shall consider a balanced draft system, since that is customary in modern steam generator practice. In this we have a supply of air at a plenum and an exhaust system at a vacuum, the latter being provided by chimney, fan, or both. We have .also the supply of fuel to regulate, either through stoker drive, feeder control, or burner adjustment. Fig. 5.29 shows the elements of the problem. The controllers may be pneumatically, hydraulically, or electrically operated; however, it is not necessary to look into the details at this point. It is sufficient if one understands that the controller is permanently connected to some power source and, in addition, has a signal input that will trigger the power at the proper time and effect a change in position of the damper, the stoker motor controller, etc. The steam pressure variation which follows load change is converted into the proper signal to regulate fuel feed and air supply. These increase or decrease properly so that a suitable new rate of combustion is produced. The manner in which the apparatus affects proper air-fuel ratio is given treatment further along. Had the boiler load increased, a diminishing steam pressure, acting through a "master regulator" or "master sender," would have signaled a new and increased rate of fuel and air flow, thus bolstering up the rate of steam generation and preventing further pressure decay.

The increased air flow into the unit would increase the gas pressure in the furnace slightly, but only until the furnace pressure regulator sensed it and responded by increasing the rate of flue gas flow to the chimney. Equally practical would be a system where the induced draft control received the signal from the master regulator and the forced draft flow was controlled by furnace pressure.

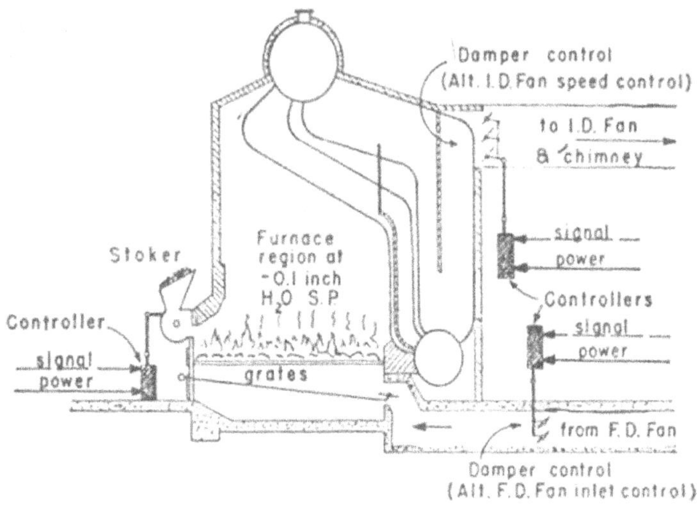

Fig. 5.29: Elements of Full Combustion Control

5.10.6. Air-fuel Ratio Control

At the present time there are three major classifications of this control. Timing, metering, and positioning control.

i) Timing Control

As is implied in the name, the combustion equipment would not operate continuously. It is the on-off system. When it is on, the fuel feed has one rate only, likewise the air flow, but, a favorable ratio can be set and maintained between them. The control starts and stops motors, opens and closes dampers, either simultaneously' or in some desirable sequence. This is especially necessary with oil and gas burners, where air flow should start and purge the combustion chamber before fuel valves are opened. The rate of combustion produced is equal to the needs at maximum rating on the 'boiler; consequently at part load there is' an excess of heat which builds up steam pressure until an upper limit of action is reached in the control apparatus, which then shuts down the combustion equipment. When the pressure has dropped to the lower control limit, the equipment is again started and the cycle repeats itself. This system must needs operate with It greater steam pressure range than the continuously modulating types to be described, else it would be cycling too frequently.

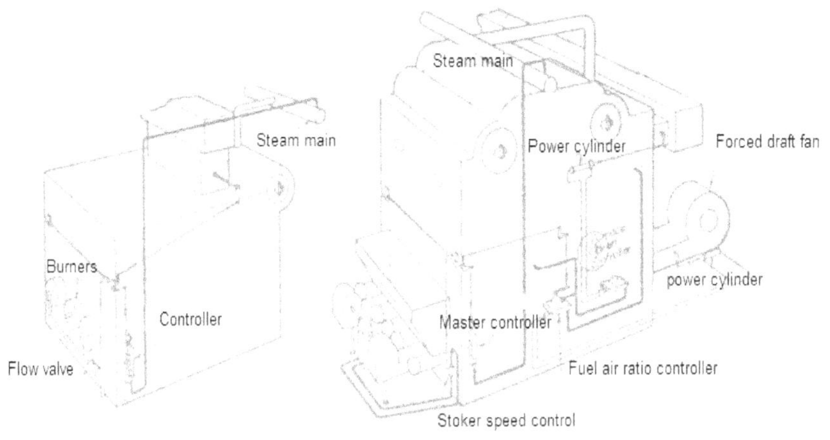

Fig. 5.30: Typical Control Arrangements

It is limited in usefulness to small plants where the auxiliary motors it starts are not large and consequently do not disturb the electrical system or strain the mechanical by their frequent across-the-line starts. Also, the cyclic temperature variations in the steam generator are undesirable in larger units. However, it is inexpensive; also it is desirable where modulation of fuel flow to light loads produces inferior combustion, as in the case with pressure atomizing oil burners. Sometimes a hybrid control is practiced, viz., modulation for the upper ranges of load, timing for the light loads.

ii) Metering Control

In this system the operation of the controller directly balances controlled magnitude (duct pressure, draft loss, etc.) against primary control impulse received from a master regulator. In metering combustion control systems the control Impulse is generally taken from the draft loss produced by the flow of gases across the boiler or across a pass, of the boiler. Thus in an air flow regulator of this type, the master, sender would cause displacement of the primary element of the controller, resulting in a damper. Change which sends in more air to the furnace. The increased gas flow through the boiler then produces more pressure difference in the draft leads back to the controller, which in turn brings the controller into balance at the new operating point.

iii) Positioning Control

This is also' called "compensated control" In this system the; controlled apparatus, such, as fan rheostat or damper lever, is brought; to definite predetermined positions established by calibration to correspond ''to the boiler output required by the control impulse. The calibration is accomplished during the initial phases of operation of the control system. The proper positions of equipment for optimum air-fuel ratio at each load are determined by manual operations and the controllers are adjusted so, that they will assume these positions as shown in figure 5.30.

In the simple mechanical variant of this system the movements of air flow damper and fuel feed control are positioned by one controller which responds to signals from the master sender. The single power unit operates both regulating points with mechanical linkages such as cable and sheave, shaft and lever, etc. The other variation is remote positioning control consisting of individual power units responding to impulses from the steam pressure controller. Many different successful schemes of control have been placed in service by the several firms specializing in combustion control systems. The following outlines' some of the possibilities.

a. Control stoker motor, forced draft fan, and induced draft fan speed from master regulator. Furnace pressure regulator to operate on boiler outlet damper.

b. Control pulverizer feeder and induced draft fan from master regulator. Furnace pressure regulated by control on forced draft fan speed or damper position.

c. Control boiler outlet damper by master regulator (steam pressure). Control stoker speed by air-flow-steam flow relation. Furnace pressure regulated by varying forced draft, through fan outlet damper, from furnace pressure controller.

d. Control induced draft by damper actuated from steam pressure master controller. Control furnace pressure by regulator which operates forced draft fan inlet vane position. Use gas flow, as metered by pressure drop across boiler pass, to control coal feed.

5.11. Feed Water Control

5.11.1. Basic Requirements

To maintain correct level of water in the drum of a large high-pressure boiler requires close control of three main functions involved in steam generation, i.e. water level, steam flow, and feed water flow. The correct measurement of the true level of water, and the correct direction of control action to be applied during changes in load, is essential with very large boilers mainly because of the relatively small drum capacity which requires quick-acting accurate control to keep water levels within gauge limits on varying load.

Control action must also take into account the effects of "swell" and "shrinkage" of water level in the drum. "Swell" is due to the expansion of the submerged steam volume in the steam-generation tubes of the boiler, which is caused by the fall in pressure when the demand for steam increases due to a rise in loading on the turbo-generator. Likewise, "shrinkage" takes place when reducing load because the boiler drum pressure increases momentarily. Therefore, when load increases, the drum water-level actually increases momentarily, and the signal from the water-level gauge, although correct in itself, causes a reduction in water flow which is the wrong direction of control action.

A change in load on a boiler-turbine unit also causes changes in steam temperature as the flow rate through the super heaters and reheaters changes, as described in the previous sub-section, Feed water flow therefore has an interaction effect on combustion control and fuel firing rate, and the tendency of "swinging" between these controls has to be considered when setting up the whole control system,

5.11.2. Feed Water Control System

To facilitate the explanation of a feed water control system it may be useful to refer firstly to the three-element pneumatic control arrangement used on many medium-sized boilers in existing power stations, and indeed also in industrial applications. A typical control scheme is shown in Figure 5.31.

Water-level is measured by a temperature equalizing column, while steam and water flow are obtained from orifice plates, The differential pressures from the steam and feed water devices are converted into pneumatic signals, which are compared, and the ratio between them transmitted to the standardizing relay. This relay also receives a pneumatic signal from the level indicator, and the combined pneumatic signal is applied to the feed water regulating valve through an auto-manual station. An auto-manual station simply means that .if any of the control elements are not functioning, or if the flows are outside the normal range, as, for example, during start-up, the operator can take over the adjustments to the regulator.

Figure 5.32 shows a diagram for the automatic feed water control of a large boiler having twin steam mains with both running and standby variable-speed feed water pumps. It can be seen that the ratio of steam and water flow measurements is applied, in this case by means of a beam balance to another beam which also receives a force proportional to drum water level.

In the more modern control schemes, such as for 500 MW boilers, two other signals are required. One is a drum level correcting signal which ensures that compensation is provided so that the drum level signal is corrected for pressure variations which cause density changes as already mentioned. The second signal is obtained from the steam flow measurement which is then characterized and summated with the drum level signal. The effect of characterizing is to demand a higher level when load demand increases and lower level for load decrease, thus compensating for "swell" and "shrinkage". '

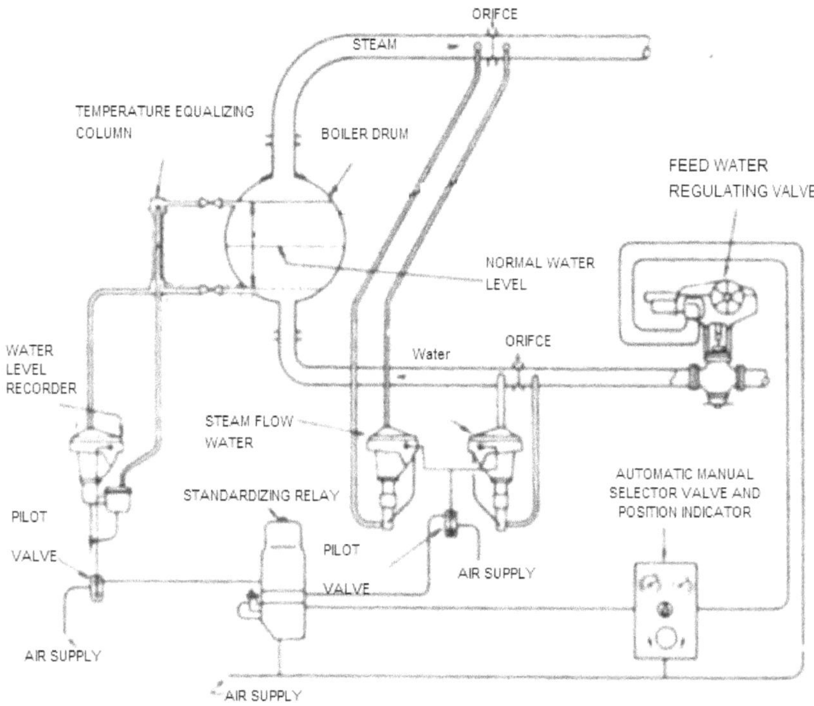

Fig. 5.31: Diagram of Three-Element Feed water Control (Air-Operated Type)

5.11.3. Computer-initiated Feed Water Control System

The operation of the control system is initiated by either a sequence controller or a computer, depending on the particular overall concept adopted. Any reference to an operation being "programmed" means that the operation is initiated at the correct time by a sequence controller or computer; but on failure of these, manual operation can be used.

From start-up to about 20% CMR the start-up feed water regulating valve position is determined by the single-element drum level controller LC_1. The set point of LC_1 is lower than the normal set point of the three-element controller LC_2, to avoid interaction between the respective control loops. The speed of the start-up pump is determined by the pressure drop controller ΔPC_1 so as to maintain the required pressure drop across the feed-regulating valves. ΔPC_1 is programmed to be biased during start-up so as to ensure an adequate flow through the start-up feed-regulating valves. This counters the effect of starting the circulation on an assisted circulation boiler, as the water level may fall several inches. During starting-up the higher drum level set point of LC_2 is programmed to be biased lower than LC_1 so as to keep the main feed-regulating valves closed.

Whilst the start-up pump is in operation, a positive bias is programmed to the pressure controller ΔPC_2, thus setting the main feed pump turbine speed governor to maximum setting. When sufficient steam is available it is admitted to the main feed pump turbine, and the main feed pump commences to deliver water. When, due to increase of load, the supply of bled steam exceeds the demand, the main feed pump turbine speed governor will control the bled-steam turbine at maximum setting.

Due to the above-mentioned bias on ΔPC_2 it will override the action of ΔPC_1 (at its normal setting). The main feed pump, will, therefore, progressively take over the supplying of feed water, the start-up pumps progressively reducing speed until they cease to deliver feed water. At this time the start-up pumps are shut down and the programmed bias removed from pressure drop controller ΔPC_2. The feed water is now being supplied by the main feed pump, and the water-level is being controlled by means of the main feed-regulating valves under the influence of the three-term controller LC_2.

It is considered that the variation of the main feed pump speed is the most desirable method of regulating feed flow, and this method is adopted when the main feed pump has taken over the duty of supplying all the required feed water. The changeover is accomplished by transferring the control action of LC_2 from the feed-regulating valves to the speed governor setting of the main feed pump turbine. The method of transfer is by a programmed action of biasing unit B_1 which applies a positive bias to the signal going to the feed-regulating valves causing them to open wide, and simultaneously reduce the signal going to the lower signal selector (ΔV). Thus the speed governor setting is transferred from the pressure drop controlled ΔPC_2 to the three-element controller LC_2.

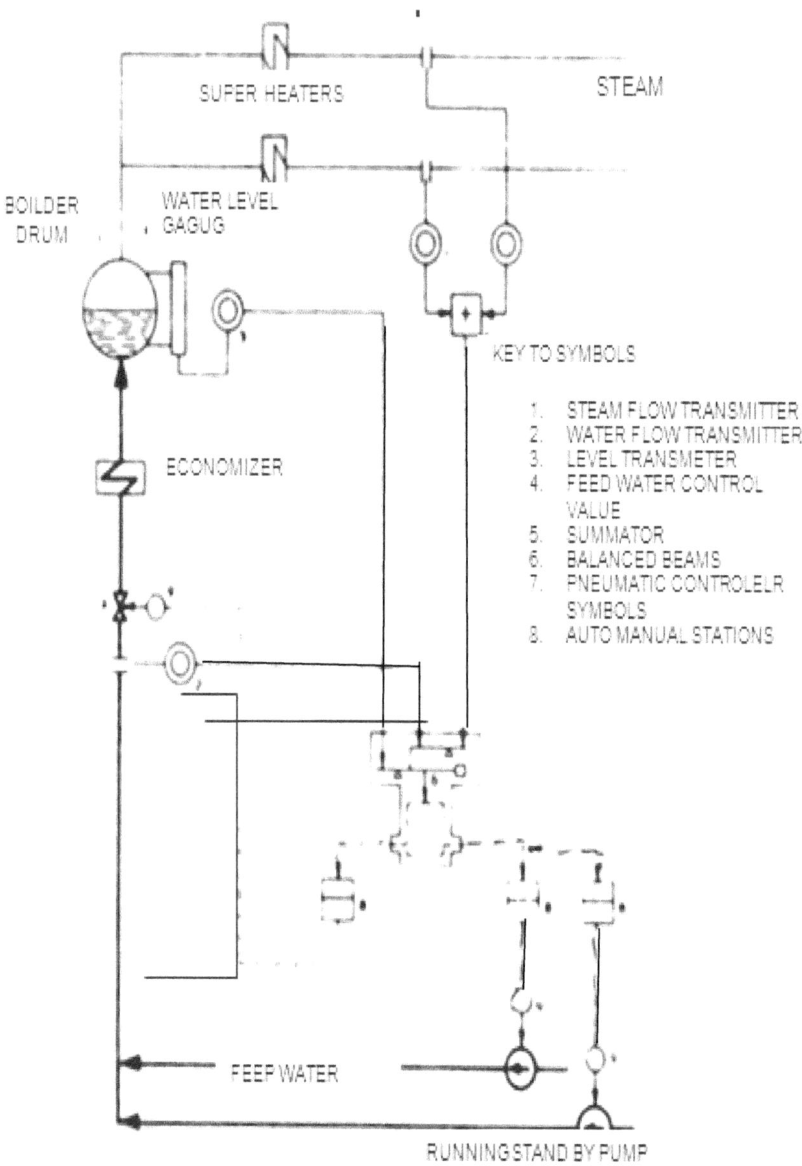

Fig. 5.32: Diagram of Three-Element Feed Water Control

(Electrically Operated Type)

If the main feed pump turbine should not respond to a reduction in demand, controller LC_2 would operate the feed-regulating valves at the biased-up value, and a similar action would result if the standby pumps started on emergency operation.

When shutting down, the above-mentioned programmed bias is removed from the control signal, from LC_2 going to the feed-regulating valves and restores the signal going to the signal selector to normal. Thus controller LC_2 transfers its action to the feed-regulating valves and the main feed pump turbine speed governor setting is again obtained from the pressure drop controller ΔPC_2.

At a programmed load level the start-up pump is brought into service simultaneously with a positive bias being applied to the pressure drop controller ΔPC_1. As the bled-steam supply to the main feed pump turbine falls, the main feed pump delivery falls off causing controller ΔPC_2 to become "saturated" when the steam-admission valves to the feed pump turbine are wide open. As the feed supply is thus falling off, the start-up pump increases its speed under the action of its control loop $(\Delta PT\text{-}\Delta PC_1)$.

When the main feed pump ceases to deliver water the steam-admission valves are closed, but the control of feed water is still under the influence of LC_2 and the main feed-regulating valves.

At a predetermined load the set point of LC_2 is programmed to be biased below the set point of LC_1. This lower set point will cause LC_2 to close the main regulating valves. At the correct water-level LC_1 will cause the start-up feed-regulating valves to open. The remainder of the shutting-down procedure for feed water control is now under the influence of LC_1 controlling the 20 % duty feed-regulating valves.

6. COMMISSIONING AND TESTING OF BOILER

6.1. Introduction

Commissioning of a boiler will involve putting the boiler into regular service and proving its capability of generating the specified pressure and temperature parameters reliably for the specified quality of fuel and water. The modern boiler has a large number of sub-system has a large number of auxiliaries. Thus it is a perquisite to the commissioning of the boiler that the various auxiliaries are tested and trial running completed. Besides its own auxiliaries the boiler requires the various service like coal handling plant, oil handling plan, water, treatment plant, ash treatment plant, ash handling plant., instrumentation and control system, LT and HT switchgear, cooling water, lubricants etc. which are required to be made available at various stages of commissioning. The commissioning of boiler can be divided into following stages:

1. Preparation of commissioning procedures
2. Inspection of boiler
3. Trail runs of equipments
2. Pre-commissioning tests
3. Commissioning of boiler
4. Trial run of the boiler
5. Efficiency/performance test of the boiler

6.2. Preparation of Commissioning Procedures

The commissioning of boiler is required to be completed in a certain time-frame and be dovetailed to the unit commission schedule. Thus it will not be exaggeration to say that time is the main guiding factor of the commissioning. As such the sequence of the various activities would have to be carefully throughout, schedules made and personnel trained. The training aspect is of car amount importance as the personnel involved in commissioning are required to have full knowledge of the intricate details of the boiler and its auxiliaries not only the 'how" but also the 'why' . They are also required to guide the operations and in many cases take over the operations themselves.

6.3. Inspection of the Boiler

This is the first important activity of the commissioning engineer at site and should cover the main boiler; its sub-systems and auxiliaries. All parts of the boiler are to be inspected thoroughly for any incomplete work. The air and gas paths are to be checked for completion work and cleared anywhere in the system. The boiler drum is to be checked internally for proper assembly and fittings of the drum internals and cleared of all the unwanted materials.

The equipment checking will have to cover proper assembly, mounting and alignment. Dampers and valves are inspected for their trouble-free operation and the open and close position limit switches properly set. Hanger supports of piping are to be inspected and ensure that the pipes are free to expand.

6.4. Trial Runs of Equipments

All the rotating equipments are to be put on trial run individually and run for a predetermined time. It is customary to run the large motors for 8 hours depending upon the supplier's recommendations. During the period of trial running the bearing temperature and vibrations are constantly monitored and periodical (half-hourly) readings recorded. A watch is kept on the equipment for any unusual noise. It is at this stage that the interlocks and protections of the individual equipment and its associated sub-systems/auxiliaries are also tested and established. This phase can also be used to impart in training to the plant operators.

6.5. Precommissioning Tests

Precommissioning tests which are carried out after the successful completion of trial running constitute of the following.

1. Gas tightness test
2. Air tightness test
3. Hydraulic test
4. chemical cleaning of boiler
5. Drying of refractory
6. steam blowing
7. safety valve setting
8. testing of protections and interlocks

6.5.1. Gas Tightness Test

The aim of this test is to check and ensure that no there is no air- infiltration into the gas system at the entire flange joints, expansion joints etc. After closing all the man holes and observation ports in the furnace and in the gas path etc. all the dampers in the air path and fuel path leading to combustion chamber are closed. The ID fan is then started and the design draft is maintained in the furnace. Air infiltration is checked at all the vulnerable points by means of lighted candles or torch whose flame would be drawn towards infiltration point. After rectifying the points of infiltration, the test is repeated till it is ensured that the system is healthy.

6.5.2. Air Tightness Test

This test is carried out by pressuring the air system by running FD. Fans. Leaks are detected at the probable points by using lighted candles and applying soap solution. Some bombs are used for detection of leaks in ducts in USA. All leaks are rectified and the test repeated until the system is healthy.

6.5.3. Hydraulic Testing

All pressure parts, after installation, are hydraulically tested as per IBR code at a test pressure of 1.5 times the maximum allowable working pressure. Before conducting the hydraulic test, the following inspections are made:

- All debris removed manually
- All welds completed, X-rayed and stress relieved
- All valve installed and closed
- All safety valves installed and hydrostatic plugs assembled
- All drains, vents and test valves open
- All manhole closure plates and header manhole caps closed and bolted.
- All instrument tapping sealed.

Either the same pump or different pumps may be used for filling and pressuring. Prior to filing, the pressure parts can be tested with air free from dust and oil and available at maximum $2Kg/cm^2$. This would aid in locating any cuts or gouges not observed by visual inspections. The demateialized water used for hydraulic test should be of good quality such as pH-7.5 to 9.0 at 25ºC, conductivity less than 1 micro ohm/cm at 25ºC , silica less than 0.02 ppm and having no oil and chloride and nil hardness. Treated condensate containing 10ppm of ammonia and 300ppm of hydrazine and pH value of 10 may also be used. Water temperature for test should not be less than 21ºC .The pressure readings during test are taken from the pressure gauge installed at the highest point of the boiler.

During testing, first the pressure is raised to 80% of design pressure when the plugs are inspected and tightened. The pressure is then carefully raised to test pressure ensuring it does exceed test value by more than 6% at any time. It is then reduced to not less than maximum allowable working pressure and maintained to examine the boiler for leaks. If leaks are observed, it is necessary to release the pressure. The leaked areas are drained, repaired, refilled and tested again. After test, it is essential to protect internal surfaces from oxidation which may be done by filling the unit writhe hydrazine and 10 pp ammonia. Nitrogen may then introduce form drum vents to pressurize the unit to 0.3 to 0.5 Kg/cm^2 or the complete system may be kept pressurized to 3 to 5Kg/cm^2. This test is once carried out by the erection personnel and is repeated as part of pre-commissioning test.

6.5.4. Chemical Cleaning

It is very important to maintain clean heat transfer surface in boiler. The mechanical cleaning of entire boiler, due to large number of long inaccessible circuits, is impossible. Chemical cleaning removes coating of iron oxide formed during fabrication or hot working process or during erection etc. If these oxides are not removed, these may be eroded and redeposited in critical areas causing local overheating and associated corrosion problems. Excepting super heater and reheated tubes, all other parts of boiler are cleaned by the process comprising of following steps.

- Water washing
- Pre-boil of surfaces with a strongly alkaline solution
- Soaking or circulation of an acid solvent with suitable inhibitor
- Neutralization with a strong alkaline solution
- Passivation of the freshly cleaned surface with ammoniated hydrazine solution.

6.5.5. Water Washing

It removes loose dirt and water soluble ingredients. The process consists of filling boiler through economizer up to normal level and draining and repeating it till practically clear water is observed in drain.

6.5.6. Alkali Boils Out

It removes preservative, grease, lubricants and temporary coating and also loosens mill scale which will otherwise render the subsequent acid cleaning ineffective. The concentration of chemicals for alkali boil out is: Hydrazine hydrate -200ppm or 0.02% and tri sodium phosphate–1000ppm or 0.1%. For alkali boil out the boiler is filled with DM water chemicals added from drum man hole and closed, boiler lighted up and pressure raised gradually to 40atg and held for 16-20 hours, with hourly blow downs for one minute each time to remove substantial quantities of impurities collected in the water wall headers. Feeding rate is maintained to keep a normal level in the drum. Phosphate level in drum should he maintained at 1000ppm by phosphate dosing. Sample are taken every two hours and analyzed for pH alkalinity, oil, PO$_4$ conductivity and silica. Blowing operation is continued till oil content is zero. After that system is cooled and drained completely at pressure of 1.5Kg/cm^2 gauge. The whole system is then rinsed using DM water till phosphate contents are nil and drain samples are close to quality of DM water. During chemical cleaning boiler internals are removed. After alkali boil out, the drum manholes are opened and any loose accumulated material in the drum is removed manually.

6.5.7. Acid Cleaning

It is cone by using dilute inhibited (4% concentrated) hydrochloric acid adopting circulation method. Cleaning by circulation method is aided mechanically by the constant change of solution, It is important that the effectiveness of the inhibitor contained in the acid be verified before admitting any acid into the boiler.

6.5.8. Nitrogen Capping

It is adopted to protect the freshly cleaned metal surface from being corroded due to exposure to atmospheric air during rinsing stages of acid cleaning operators. Nitrogen capping may be terminated only after the completion of sodium carbonate wash.

6.5.9. Soda Ash Washing

It is done with circulation of mild sodium carbonate (1.0%) solution at an elevated temperature to prevent spotting on freshly pickled surface. Temperature is raised to 80ºC by addition of steam during circulation, and circulation maintained for 4 hours. In between acid cleaning and final naturalization with soda ash washing, the three operations of DM eater rinse, citric acid rinse and final rinse with DM water are also carried out.

6.5.10. Drying Refractory

The refractory around burner throats, some portions in the combustion chamber of furnace and roof may contain some moisture after erection. It is necessary to remove this moisture so that the refractory finally sets in properly to avoid any cracks or buckling on the brick work. This it achieved by slow heating of the brick work uniformly. This is partly achieved during the chemical cleaning. It is completed after lighting up by maintaining uniform temperature.

6.5.11. Steam Blowing

Steam blowing operation is carried out before passing steam to turbine in order to remove scale, loose material iron cuttings etc. Which might have been entrapped in main steam pipes, super heaters etc. during manufacture, storage or erection. If these are not removed initially, then the turbine blades may be damaged during normal operation.

The scale etc. deposited on pipes is first dislodged by thermal shock and then flushed out by the steam medium, the method being known as a puffing method. The boiler pressure is raised to a predetermined value, and then the firing shut off and at the same time the valve in the pipe line opened. In order to avoid undue thermal stresses in the boiler drum, the pressure drop in drum is limited to the corresponding temperature change of around 40ºC . To achieve such condition the drum pressures after blowing should be 3.5,7,13,18.5,25 and 31Kg/cm^2 a initial blow down pressures of 10,20,30,40,50 and 60 Kg/cm^2 respectively. After this much pressure drop, the valve is again closed and firing re-started. About 20 number of blow off operations requiring 2 days time may be required to have completed cleaning of the pipes. The idea of the debris being removed from the pipe can be had from the colour of the steam discharged to atmosphere. The effectiveness of the blowing can be determined by noting abrasions on the polished mild steel target plates attached to the end of exhaust piping. For the final or last two blowing, the target plates of same material as turbine blades are used in order to have estimate of erosion during regular operation.

The operation of blowing requires huge quantities of dematerialized water which for a typical unit of 200MW may be around 6000M^3. It is also essential to carry out internal boiler water treatment during blowing operation with addition of ammonia or hydrazine so as to maintain a pH value of 10 in boiler water and in feed water 5 to 25 ppm of hydrazine.

6.5.12. *Safety Valve Setting*

The safety valves are required to be floated and set to operate at the desired set pressure. It is a statutory requirement that this is done in the presence of the boiler inspector. The pressure is gradually raised above the normal working pressure up to the set pressures of the safety-valves. The safety valve with highest set points is floated and set first and the lower one nest and so on. During the time the valve with higher set point is set, the valves with lower set points are incapacitated by blocking their operation (gagged).

6.5.13. *Testing of Protections and Interlocks*

As already indicated the protections and interlocks are tested and established in so far as these pertain to the auxiliaries before the latter are being trial run by simulation tests where actual tests are not possible.

The boiler protections and interlocks as such are once again tested by simulation where necessary repeatedly. One of the systems that is required to be carefully gone over is the furnace safeguard surveillance system incorporating the main fuel trip relay (MFR).

6.6. Commissioning of Boiler

After completing all the pre-commissioning tests satisfactorily, the boiler is ready for regular commissioning and putting it into regular operation. This is in so far as the boiler unit is concerned. For commissioning of the boiler, however the turbine-generator is required to be ready and the pre commissioning test on it should have been completed. Also the instrumentation and control systems should be complete in all respects and should have been established.

6.7. Trial Run of the Boiler

Before trial running a boiler it is commissioned and the automatic controls of furnace draft, drum level, steam temperature and combustion control are commissioned. The trial run, which is undertaken to proven the boiler's capability is conducted as agreed upon is the contract. The total   period of running is normally 14 days at varying loads with about 48 to 72 hours of continuous running on economical load an 24 hours of full load at the specified parameters of steam and feed water, at the end of these trial runs, the boiler is declared to be ready for commercial operation.

6.8. Testing of Boiler

The testing of a boiler involves the checking of the following:

- Nominal output of the boiler(steaming capacity)
- Nominal pressure of the superheated steam
- Nominal temperature of the superheated steam.
- Efficiency at maximum continuous rating for an average ambient temperature.

Besides the above, the temperature and drafts of the flue gas in the combustion chamber at heat recovery zones; and temperature of feed water and air are also measured and compared with the designed values. It is essential for checking the guarantees furnished by the supplier that the fuel and the feed water used are of the same analysis as specified during placing order and according to which the boiler has been designed. It is the usual practice that before carrying out the guarantee test, first a verification test be conducted, the purpose of which is to familiarize all the participants with their respective function and also to verify the participants with their respective functions and also to verify the accuracy of the installed instruments. Depending on the results of the verification test, all the mistakes detected may be rectified and then guarantee test conducted.

Prerequisites for Conducting Test

- The outside heating surface should be practically clean.
- All such equipment which is exposed to wear and tear, such as vanes of ID and FD fans, balls in mills and other rotating parts must be inspected thoroughly.
- All the forms which are necessary for entering the measured values should be ready.
- All the required instruments should be properly installed and calibrated and their correction curves prepared.
- All the arrangements necessary for collecting and weighing of samples and conducting their analysis should be ready.
- The orifice plates and flow meters should be checked.

6.9. Programme for Carrying Out Guarantee Test on the Boiler

The guarantee test has to be so programmed as to carry it out continuously without any break. Normally the programme is for six days as under and may be extended depending on the site conditions.

First Day

Internal and external inspection of the boiler is carried out and the burners and control devices checked. Measuring equipment is installed and the orifice plates/nozzles and flow meters for steam and water are checked.

Second Day

The first day's work is finished and if possible, the boiler is lighted up also.

Third Day

Boiler is operated at the nominal output and on automatic regulation to obtain steady condition. The attendants are issued instruction regarding their respective job and verification test is conducted for duration for four hours.

Fourth Day

The analysis of the samples taken during the verification test is made and the evaluation of the results is conducted.

Fifth Day

If everything is normal, the actual guarantee test is started. The duration is generally 4 hours. During the period when guarantee test is conducted, continuous blow-down is kept closed.

Sixth Day

The test instruments are disconnected and the preliminary calculations made and the report of the test is prepared as shown in the table 6.1.

Table 6.1: Frequency and External of Measurements

Measurement	Frequency
Samples of raw coal before mill	15 mts
Quantity of steam flow	3 mts
Pressure and temperature of steam	5 mts
Quantity of feed water flow	3 mts
Pressure and temperature of feed water	10 mts
Temperature and draft of the flue gas	10 mts
Analysis of flue gas	15 mts
Ambient temperature and temperature at the suction of fans	15 mts
Hot air temperature and pressure	10 mts
Atmospheric pressure and humidity	At beginning and end
Ash sample of fixed residue	15 mts
All readings of operational instruments mounted on the board	15 mts

In addition, separate instruments are also installed for accurate measurement of steam and feed water flow, temperature and pressure of superheated steam feed water and flue gas. The temperatures of feed water before and after economizer, of steam before and after the injector, of flue gas after super heaters, before and after economizers and air heaters and at boiler outlet and in combustion chamber, of air at fan delivery and after air preheaters are also measured. The other important parameters are pressures of air at forced fan delivery, after air heaters, total air flow, primary air flow, and temperature of flue gas before mill, draft before and after mills, draft and temperature after classifier.

6.10. Efficiency Evaluation

The efficiency of boiler is generally calculated by indirect method, in which the following losses are computed and it is pressured that rest of the energy is utilized usefully in the boiler. The various possible sources of heat loss of the boiler are:

1. Loss due to mechanical imperfect combustion
2. Loss due to physical heat carried away in slag and fly ash.
3. Loss due to imperfect combustion or chemical loss.
4. Loss due to chimney gases.
5. Loss due to radiation.

6.11. Normal Boiler Starting Procedure (Cold Start-up)

The boiler needs to be brought to the normal operating pressure and temperature with heating rate not exceeding permissible metallurgical limits.

*Prepare boiler for start-up: All instrumentation and associated mechanical equipment has been checked and available for service.

Performance walks down check list like

1. Check ID, FD and SA fans, bearings, pneumatic cylinder, damper position and controls etc., make sure that the dampers are operational.
2. Check the drum level gauges and compare with DCS readings. Test the gauges if necessary.
3. Check the close/open position of all isolation valves.
4. Check that all man doors are properly closed.
5. Check all soot blowers are retracted.
6. Check the fed water pumps. Recirculation valve should be 100% open. Check for cooling water flow. Ensure the suction valves are fully open before starting the feed water pumps.
7. Ensure that the boiler is filled up to the operating level.
8. Drum vent, super heater vent & start up vent should remain open. Drum vent should be closed if the drum pressure reaches 2 Kg/cm^2.
9. Ensure that the desuperheating control and isolation valves are closed.

During the initial hours of starting it is better to keep open the super heater drain to remove completely any condensate and can be closed once steam starts coming continuously through the drain. The startup vent should remain open during the entire start up period. Prepare the boiler for fuel firing. Initial warming up with firewood should not take more than 4 hours. This is enough for our boiler refractory drying. Start ID and SA fans before starting fuel feeding FD fan staring can be done during the final stage of pressure raising. It is recommended to keep only low firing rate in the initial hours by running the feeders with reduced discharge to control the rate of temperature raise. Based on experience a drum temperature raise of 30ºC/hr is usually recommended to avoid temperature stress and uneven boiler expansion movements. The steam temperature should not be allowed to rise more than 50ºC/hr or25ºC/every half an hour during the entire start up period. The temperature during the initial hours should be strictly controlled. After the drum temp raise has reached 150ºC, focus should be switched to pressure raise, rather than temperature raise.

Recommended temperature raise is 30ºC for the fifth hour, 40ºC for the sixth hour and 50ºC for the remaining start-up hours. Restrict 50ºC differential temp at upper and lower side of the steam drum during start-up.

On Pressure Basis Raise or Pressure should be Controlled as Below

1. When pressure starts raising control pressure at the rate of $0.1Kg/cm^2$ per minute. That is $6kg/cm^2$ per hour for the fifth and sixth hour. At the end of 6 hours, steam drum pressure should be $12 Kg/cm^2$ Close drum vent at $2 Kg/cm^2$ start-up vent should remain open.

2. Continue pressure raising at the rate of $0.15Kg/cm^2$ per minute for the seventh hour.

3. Continue at the rate of $0.2Kg/cm^2$ for the eight hour during every stage of pressure raising the boiler engineer should go around the boiler checking the expansion movements, carefully observing the economizer areas, bank tube areas super heater areas for possible tube leakages.

4. Proceed further at the rate of $0.25 Kg/cm^2$ main for the ninth hour and $0.3Kg/cm^2$ for the tenth hour at the end of tenth hour, steam drum pressure will be reaching the operating pressure. Take boiler into operation at the end of tenth hour. Start up vent should be closed only after establishing enough steam flow.

7. STEAM TURBINES

7.1. Introduction

Turbines are devices that spin in the presence of a moving fluid. The difference between water wheels or windmills and turbines is largely one of emphasis and degree. During the 18th and 19th centuries, much progress was made toward extracting the kinetic energy of flowing water by devising water turbines. Leonhard Euler, applying fluid mechanics, developed a water turbine as early as 1750. During the 18th century several engineers, such as Benôit Fourneyron, succeeded in building water turbines that by far outstripped conventional water wheels by giving the blades special shapes. The term "turbine" was coined by Fourneyron's professor Claude Burdin; he derived the term from *turbo*, a spinning object.

The most useful turbines for many purposes are those that can be propelled with energy from heat. A typical turbine based on heat is the steam turbine. The idea of a steam turbine is much older than the steam engine itself. Around 60 BCE the Alexandrian Greek Heron (a.k.a. Hero) used jets of steam to turn a kettle. In 1629 the Italian engineer Giovanni Branca depicted in his machine book *Le Machine* a steam turbine in which a jet of steam is directed at the vanes of the same sort of apparatus as a water wheel. No doubt others observed that escaping steam is like the rushing wind and could be used to push mills just as the wind powers windmills. When practical steam engines were built at the start of the 18th century, however, they moved a cylinder back and forth (reciprocating motion) instead of pushing a wheel around, although they could be made to turn wheels with various ingenious mechanisms. Reciprocating steam engines were bulky, had slow rotation speeds, and wasted much energy in the machine itself to move the heavy pistons back and forth. When first used to drive electric generators, reciprocating steam engines proved difficult to maintain at a fixed rotation speed as the load on the generator changed.

Turbines are as simple as reciprocating engines are complex. Because they have essentially only one moving part, they are sometimes called the perfect engines, almost directly turning heat into rotary motion. The first to build a steam turbine was the British engineer Charles Algernon Parsons. In 1884 he completed a small turbine that rotated at 18,000 revolutions per minute and that delivered 10 horsepower. The Swedish engineer Carl Gustav de Laval, experimenting with steam turbines, achieved greater power and higher rotation rates. In 1890 he built a turbine consisting of a 30-cm (12-in.) disk with 200 blades mounted on a flexible axis. The steam was admitted to the blades by special nozzles (Laval nozzles) that accelerated the steam to very high velocities, thus transferring the energy of the steam in the form of kinetic energy to the blades. The design of steam turbines developed into a science near the end of the 19th century. Better materials allowed the construction of turbine blades that are resistant to corrosion. Charles Curtis developed the multistage turbine in which the blades and disks become progressively larger when the steam expands. Parsons developed in 1894 the ship turbine engine. The slow-revolving turbine consisted of several sections of increasing diameter. High-pressure steam is admitted to the turbine and pressure differences in each section drive the turbine blades. The first ship to be equipped with such a steam turbine, the *Turbinia,* immediately established a speed record with 31 knots (57.5 km or 35.7 mi per hour). During the early years of the 20th century, most reciprocating steam engines were replaced by steam turbines(or by diesels). Steam turbines can deliver much more power than reciprocating engines and need less maintenance. Steam turbines also supplanted marine steam engines on ships.

A similar evolution took place for large internal combustion engines, mainly driven by the need for lightweight and powerful airplane engines. Most large modern airplanes are now powered by either turboprop or turbojet engines. These turbines are spun by the expansion of jet fuel instead of by the expansion of water into steam.

7.2. Theory of Operation

A working fluid contains potential energy (pressure head) and kinetic energy (velocity head). The fluid may be compressible or non-compressible. Several physical principles are employed by turbines to collect this energy.

Impulse turbines change the direction of flow of a high velocity fluid jet. The resulting impulse spins the turbine and leaves the fluid flow with diminished kinetic energy. There is no pressure change of the fluid in the turbine blades. *Pressure head* is changed to *velocity head* by focusing the fluid with a nozzle, prior to hitting the turbine blades. Pelton wheels and de Laval turbines use this concept. Impulse turbines do not require a pressure casement around the runner, since the fluid jet is prepared by a nozzle prior to hitting the turbine. Newton's second law describes the transfer of energy for impulse turbines.

Reaction turbines develop torque by reacting to the fluid's pressure or weight. The pressure of the fluid changes as it passes through the turbine. A pressure casement is needed to contain the working fluid as it acts on the turbine runner, or the turbine must be fully immersed in the fluid flow (wind turbines). The casing contains and directs the working fluid, and for water turbines, maintains suction imparted by the draft tube. Francis turbines and most steam turbines use this concept. For compressible working fluids, multiple turbine stages may by used to efficiently harness the expanding gas. Newton's third law describes the transfer of energy for reaction turbines.

Turbine designs will use both these concepts to varying degrees whenever possible. Wind turbines use a foil to generate lift from the moving fluid and impart it to the rotor (this is a form of reaction), they also gain some energy from the impulse of the wind, by deflecting it at an angle. Cross flow turbines are designed as an impulse machine, with a nozzle, but in low head applications maintain some efficiency through reaction, like a traditional water wheel.

Classical turbine design methods were developed in the mid 19th century. Vector analysis related the fluid flow with turbine shape and rotation. Graphical calculation methods were used at first. Formulas for the basic dimensions of turbine parts are well documented, and a highly efficient machine can be reliably designed for any fluid flow condition. Some of the calculations are empirical 'rule of thumb' formulas, and others are based on classical mechanics. As with most engineering calculations, simplifying assumptions were made.

Modern turbine design carries the calculations further. Computational fluid dynamics dispenses with many of the simplifying assumptions used to derive classical formulas, and computer software facilitates optimization. These tools have led to steady improvements in turbine design over the last forty years

The primary numerical classification of a turbine is its *specific speed*. This number describes the speed of the turbine at its maximum efficiency with respect to the power and flow rate. The specific speed is derived to be independent of turbine size. Given the fluid flow conditions and the desired shaft output speed, the specific speed can be calculated and an appropriate turbine design selected.

The specific speed, along with some fundamental formulas can be used to reliably scale an existing design of known performance to a new size with corresponding performance.

7.3. Types of Turbines

1. Steam turbine
2. Gas turbine

Steam and Gas turbine are sometimes technically referred to as Joule-Brayton or Rankine cycle turbines. These thermodynamic energy conversion cycles have basically 4 stages with minor variations. Wikibook on Joule-Brayton Cycle.

Compressor stage: A working fluid (or liquid or gas) is compressed. This stage is variable in order.

Heating stage: The fluid is heated (or combusted with a fuel), which causes it to expand.

Power stage: The expanding fluid turns a turbine which generates workable energy.

Exhaust stage: The used fluid is either cooled and recycled (closed cycle) or ejected from the system (open cycle).

3. Water turbine
4. Wing turbine

Water and Wind turbines do not have a thermodynamic cycle and the working fluid is not heated: the simple pressure differential from one side of the turbine to the other causes fluid to spin the turbine.

7.4. Principle of Operation of Steam Turbine

Steam turbine converts the heat energy of steam (contain in steam by virtue of high pressure and temperature), into mechanical work. The steam turbine depends wholly on the dynamic action of steam. The components of a simple steam turbine can be seen in fig 7.1. The construction and working of the turbine are As follows: The high pressure steam is allowed to expand to a lower pressure in one or more properly shaped passages called nozzles. These nozzles are attached to the stationary part of the turbine which is called stator or casing or cylinder. The expansion of steam in the nozzle converts some of the heat energy in the steam into kinetic energy. The steam with a greater velocity now flows over a set of curved blades; the steam particles undergo a change in the direction of motion. This causes a change in its momentum. The force thus developed causes the wheel and therefore the turbine shaft to rotate. In a steam turbine unit, a fixed element (nozzle) and its corresponding rotating element (rotor wheel) form a turbine pair or turbine stage. The steam turbines used are two types, namely

1. Impulse turbine
2. Reaction turbine

7.4.1. *Impulse Turbine*

In an impulse turbine, the steam is expanded, from its initial high pressure to the lowest pressure in nozzle fixed to the casing. This high velocity steam glides over the blades fixed on the periphery of the rotor. These blades change the direction of steam flow, without changing its pressure. Due to the change in momentum motive force is caused. This force makes the turbine shaft to rotate. The principles of operation of the simple of the simple impulse turbine can be seen in Fig. 7.2. The top portion of the figure shows the longitudinal section through the upper half of the turbine, the middle portion shows the development of the nozzles and balding, while the bottom part shows the manner in which the absolute pressure and absolute velocity of steam vary from point to point during its passage through the turbine. The following features can be seen in this figure. The pressure of steam falls in the nozzle and remains almost constant in the blade ring. The velocity of steam increases in the nozzles only is called an impulse turbine. Since there is only one set of nozzle, this turbine is called simple impulse turbine.

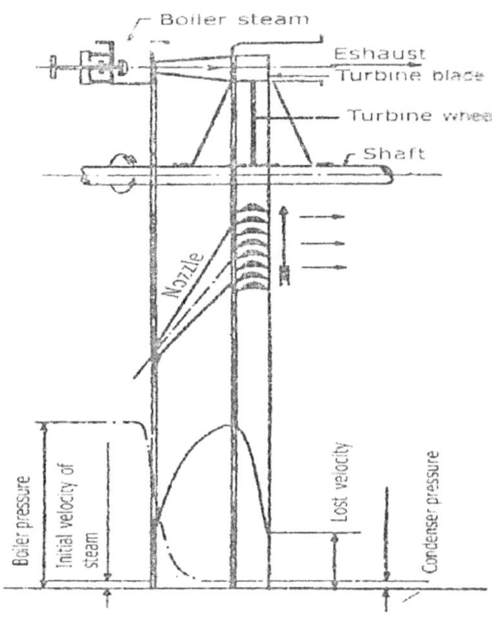

Fig. 7.1: Simple Impulse Turbine

7.4.2. Reaction Turbine

The principle of reaction can be understood by the following: When steam is discharged through a nozzle, force is required to eject the same, and a nozzle, force is required to eject the same, and consequently there is a backward reaction. Anyone who has held a fire fighting nozzle ejecting water will be aware of this tendency. No work will be done upon the nozzle if the backward movement is not allowed. If the nozzles were attached to the rim of a revolving wheel, and the fluid is discharged tangentially as shown if Fig. 7.3. Then the reaction would work in turning the wheel in the direction opposite to the jet.

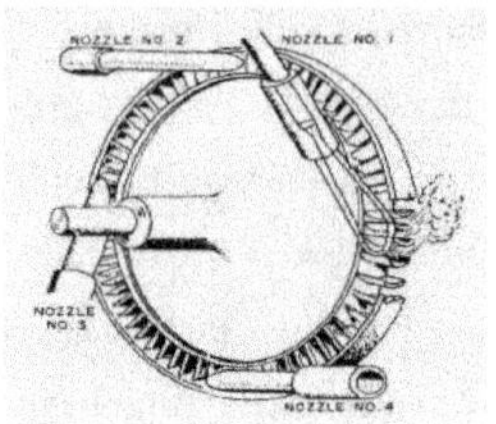 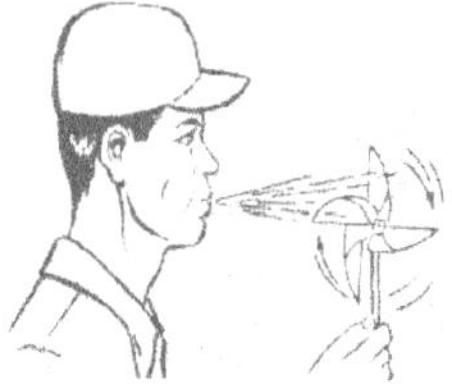

Fig. 7.2: Simple Impulse Turbine Principle

In a recent turbine, number of blade rings is fixed to the casing and the same number of moving blades is fixed to the turbine rotor periphery. The fixed and moving blades rings are placed alternately and the last one is a moving blade ring. The principle of operation of operation of a reaction turbine can be seen fig. 7.3. The top part of the figure shows the longitudinal section through the upper half of the turbine, the middle portion represents the development of the fixed and moving blades while the bottom part indicated the way in which the absolute pressure and absolute velocity of steam vary from point to point during its passage through the turbine.

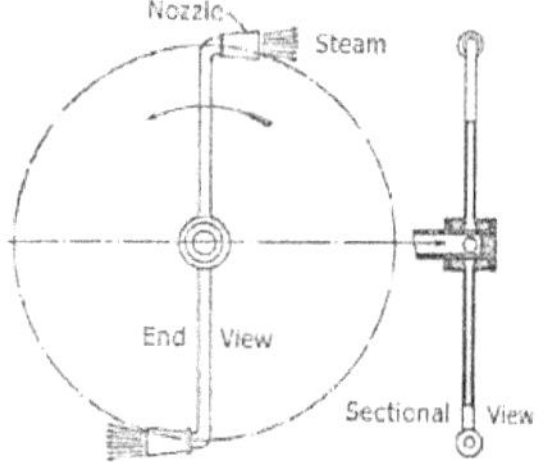

Fig. 7.3: Reaction Turbine

The following feature can be seen in this figure. The pressure of steam falls almost continuously as it passes over the blade rings. That is the steam expands both in the fixed and moving blades continuously. The velocity increases in the fixed lade ring and decreases in the moving blade ring. This type of turbine is called the reaction turbine or impulse reaction turbine. In this turbine, a set of one fixed blade ring and moving blade ring is called turbine stage. Since the above turbine, shown in the figure, contains number of fixed blade rings and moving blade rings, it is called a multistage reaction turbine. In a reaction turbine, the diameter must increase continuously as shown in the figure, in order to allow for the increasing volume of steam as the pressure decreases. There are two types of reaction turbines used in practice, namely, parson turbine and Ljung storm turbine. In the parson turbine, the steam flow is axial i.e. along the axis of the turbine shaft. In the Ljung storm turbine, the steam flow is radial. The steam passing through the moving blades determines the circumferential or peripheral force which tends to displace the rotor in the direction of steam flow. In reaction turbine, the axial forces are of considerable magnitude. Rotors are prevented from axial displacement with the aid of special devices incorporating a thrust bearing and a dummy balance piston set on the turbine shaft. Modern high capacity multistage turbine are usually impulse reaction machines, i.e the high pressure side may incorporate several purely active stages, followed by stages with a growing percentage of reaction. In multistage turbines, the overall heat drop available is realized in all stages.

7.4.3. Compound Impulse Turbine

In the simple impulse turbine shown and discussed previously, there is only one set of nozzles. The steam expands through a large pressure ratio (i.e. from inlet pressure to exit pressure) in these nozzles. As such the steam issue from the nozzle outlet with a very high velocity. When this steam jet is passed over a single blade ring or rotor, and its kinetic energy is absorbed by that ring, the resulting speed of the rotor will be too high for practical purposes. In addition, it was found that the steam leaving the rotor possesses large absolute velocity even though its pressure is low. This is means a greater loss of kinetic energy. In order to overcome high rotor speeds and to reduce the loss of kinetic energy different methods are being used. In all these methods, the energy available in the steam is converted into useful work in several blade rings. These blade rings are keyed on to a common turbine shaft in series. This methods of absorbing the velocity of the steam jet in stages is called compounding. The following are the three methods of compounding used in impulse turbines:

1. Velocity compounding
2. Pressure compounding
3. Pressure velocity compounding

In all these turbine, pressure drop occurs in the nozzle rings only and velocity drop occurs in the moving blade rings, Each disc carry the moving blades is perforated and this maintains the same pressure on both sides of the disc.

7.4.4. Velocity Compounding Impulse Turbine

The principle of velocity compounding in an impulse turbine is as shown fig 7.4. In this unit, number of moving blade rings are keyed in series on the turbine shaft. In between two moving blade rings , there is a ring of fixed blade. Just before the first moving blade ring, there is row of nozzles fixed to the turbine casing.

The steam expands in the nozzle from the boiler pressure to the condenser pressure. This produces a high velocity steam jet. This jet is passed over the first ring of moving blades. This blade ring absorbs only a portion of the kinetic energy and converts the same into mechanical work. The steam with the remaining energy is now exhausted onto the next ring of fixed blades. This fixed blade ring just changes the direction of the jet and guides the same onto the second ring of moving blades. The fixed blades do this job without appreciably reducing the velocity of the steam jet. The second moving blade ring now absorbs a further portion of the kinetic energy possessed by the steam jet. This process is repeated as the steam flows over the remaining pairs of blade rings (i.e. guide blade and moving blade rings) until practically all the kinetic energy of the steam jet has been absorbed by the blade rings.

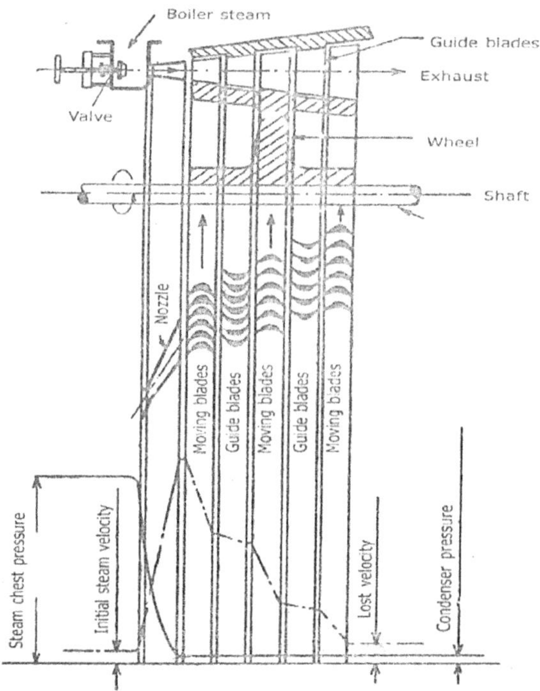

Fig. 7.4: Velocity Compounding Turbine

The arrangement of nozzles, moving and fixed blade rings can be seen in the figure shown. The variation of pressure and velocity can also be seen in the figure. The entire pressure drop takes place in the nozzle ring. The pressure remains constant as the steam flows over the moving and fixed blade rings. In this impulse turbine, since the kinetic energy is absorbed in stages by number of moving blade rings, rotor speed is lesser. This type of turbine is called impulse turbine of velocity compounded type.

7.4.5. *Pressure Compounding Impulse Turbine*

The components and the principle of operation of the pressure compounded impulse turbine are shown in fig 7.5. In this unit, number of rings of moving blades is keyed on to the turbine shaft, in series. Each moving blade ring is preceded by a ring of nozzles, fixed to the turbine casing. The steam from the boiler is passed through the first nozzle ring in which it is partially expanded.

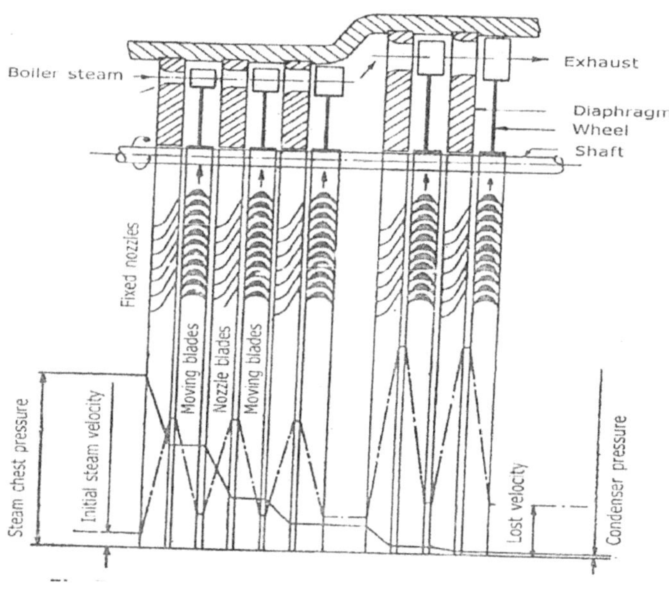

Fig. 7.5: Pressure Compounded Impulse Turbine

This increases its kinetic energy. The stem is then passed over the first moving blade ring. From this blade rings, the steam is exhausted into the next nozzle ring wherein it is again partially expanded. This expansion brings down the pressure and hence increases the kinetic energy of steam. The kinetic energy thus developed is absorbed by the next moving blade ring. This process it repeated in the subsequent rings of nozzle and moving blades, until the whole of the pressure drop has been affected. Each pressure drop affected by a nozzle ring is called a stage. Causing the pressure drop and absorbing the resulting kinetic energy in stages reduces considerably the velocity of the steam entering any moving blade ring. This in turn brings down the speed of the turbine shaft. In this figure, the pressure velocity curves show their variations as the steam flows through the turbine unit. The steam pressure remains constant during its flow over the moving blades, this type of turbine is called impulse turbine of pressure compounded type.

7.4.6. *Pressure Velocity Compounded Impulse Turbine*

The components and the principle of operation of a pressure velocity compounded impulse turbine can be seen fig 7.6. The total pressure drop of the steam is divided into stages. A ring of nozzle at the commencement of each stage causes the required pressure drop. The velocity energy thus produced in each stage is also absorbed in steps by a number of moving blade rings separated by fixed blade rings, This feature brings down the speed of the rotor. It will be seen from the curves given in the figure that the pressure is constant over the moving and fixed blade rings i.e. during each stage. As such, this turbine is called an impulse turbine of pressure velocity compounded type.

7.5. Steam Turbines-Types and Arrangements

The steam turbine used in thermal power plants may be broadly grouped into three types, the classification is made in accordance with the condition of operation of the steam on the rotor blades. The groups are as follows:

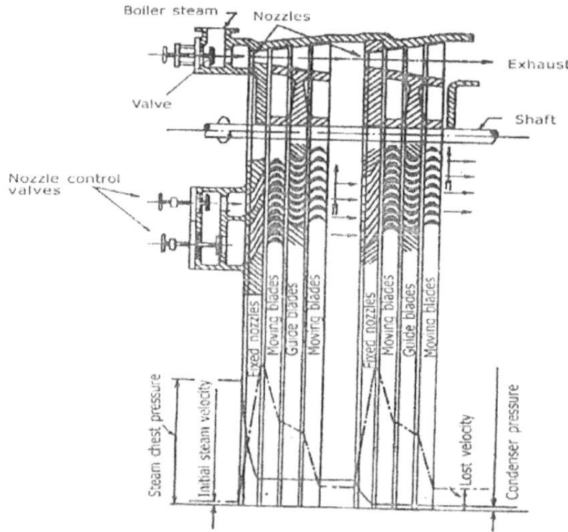

Fig. 7.6: Pressure Velocity Compounding Impulse Turbine

1. Impulse turbine-These may be divided into simple impulse and compound impulse turbine. The compounded impulse turbine can be subdivided into pressure compounded, velocity compounded, and pressure velocity compounded.

2. Reaction turbines-can be subdivided into axial flow turbines.

3. Combination of impulse and reaction turbine.

   Further classification of steam turbines can be made based on the following:

   1. Number of shafts

   2. Number of cylinders.

   3. Number of exhausts.

   2. Speed of operation.

   3. Condensing or non condensing.

   4. Extraction (bleed) or straight expansion.

Condensing or non condensing turbine are those in which the back pressure or exit pressure of steam is below or above atmospheric pressure respectively. Exhaust steam is condensed in the former type. In bleeder or extraction turbines, parts of the steam leave the casing before the exhaust. Removing small quantity of steam for feed water heating is usually called bleeding.

Non Automatic and automatic extraction turbines used in industrial plants divert major parts of the throttle steam at intermediate stages of the turbine for process uses. When waster or by product steam at reduced pressure is available, it may be led into the lower stage of a turbine to do work; this is a mixed pressure turbine. Extraction –induction units both remove and introduce steam into the turbine after the turbine after the throttle valve. The reheat turbine working on a reheat cycle returns its steam after partial expansion to the steam generator for resupeheating and then expands it to exhaust pressure. Superposed or topping turbines are high pressure noncondensing units installed in the existing plants to exhaust into the existing low pressure turbines. They increase plant capacity and improve overall performance.

7.6. Turbines for Cogeneration

Steam Turbine: Steam turbines are the most commonly employed prime movers for cogeneration applications. In the steam turbine, the incoming high pressure steam is expanded to a lower pressure level, converting the thermal energy of high pressure steam to kinetic energy through nozzles and then to mechanical power through rotating blades.

Back Pressure turbine: In this type steam enters the turbine chamber at High Pressure and expands to Low or Medium Pressure. Enthalpy difference is used for generating power/ work. Depending on the pressure (or temperature) levels at which process steam is required, backpressure steam turbines can have different configurations as shown in Figure 7.7.

In extraction and double extraction backpressure turbines, some amount of steam is extracted from the turbine after being expanded to a certain pressure level. The extracted steam meets the heat demands at pressure levels higher than the exhaust pressure of the steam turbine. The efficiency of a backpressure steam turbine cogeneration system is the highest. In cases where 100 per cent backpressure exhaust steam is used, the only inefficiencies are gear drive and electric generator losses, and the inefficiency of steam generation. Therefore, with an efficient boiler, the overall thermal efficiency of the system could reach as much as 90 per cent.

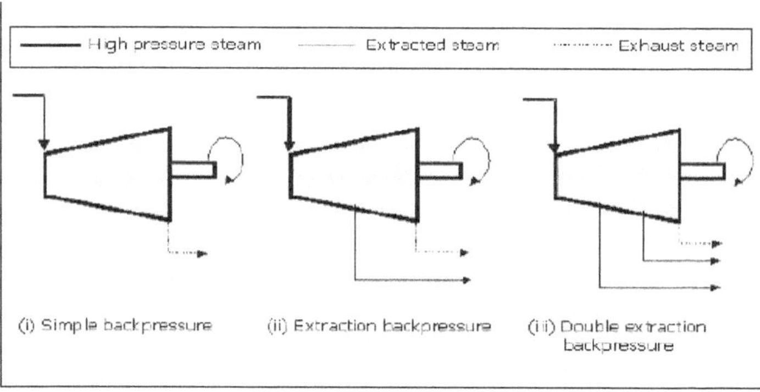

Fig. 7.7: Back Pressure Turbine

Extraction Condensing turbine: In this type, steam entering at High/Medium Pressure is extracted at an intermediate pressure in the turbine for process use while the remaining steam continues to expand and condenses in a surface condenser and work is done till it reaches the Condensing pressure (vacuum). In Extraction cum Condensing steam turbine as shown in Figure 7.8, high Pressure steam enters the turbine and passes out from the turbine chamber in stages. In a two stage extraction cum condensing turbine MP steam and LP steam pass out to meet the process needs. Balance quantity condenses in the surface condenser. The Energy difference is used for generating Power. This configuration meets the heat-power requirement of the process. The extraction condensing turbines have higher power to heat ratio in comparison with backpressure turbines. Although condensing systems need more auxiliary equipment such as the condenser and cooling towers, better matching of electrical power and heat demand can be obtained where electricity demand is much higher than the steam demand and the load patterns are highly fluctuating. The overall thermal efficiency of an extraction condensing turbine cogeneration system is lower than that of back pressure turbine system; basically because the exhaust heat cannot be utilized (it is normally lost in the cooling water circuit). However, extraction condensing cogeneration systems have higher electricity generation efficiencies

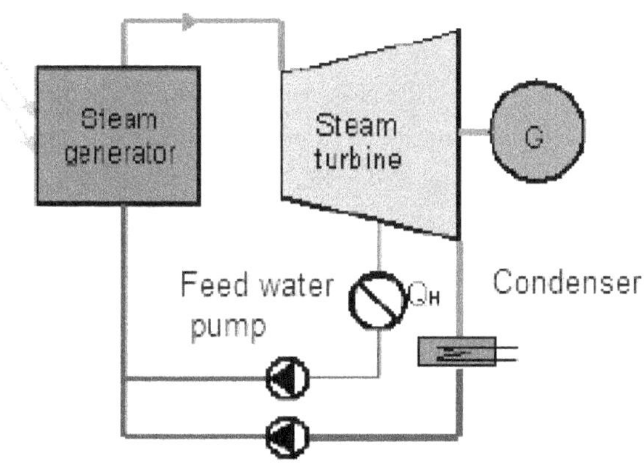

Fig. 7.8: Extraction Condensing Turbine

7.7. Construction Details of Turbine Components

High pressure casings are constructed of cast steel, the low pressure and exhaust casings are of high grade cast iron, or in certain cases of welded steel construction. All casing are divided horizontally so that the rotors can be lifted out for inspection or cleaning without disturbing the alignment of the bearings. Removable guides are provided to prevent injury to the blades when lifting off and replacing a top half casing. The joints between the two halves are plain faced metal to metal machined and scraped. The casings are lagged with heat resisting material and covered with stainless steel sheets. Steam chests are made of cast steel similar to the high pressure casings. Corrugated interconnecting loop pipes are fitted between the steam chest and the high pressure cylinder. Steam strainers are incorporated in the steam chest and adequate drains are provided. The steam chest accommodates the controlling valves which adjust the steam supply to suit the load conditions. Turbine shafts are made of high grade carbon steel of adequate proportions to ensure safety under any abnormal over loads. Air hardened steel are chiefly used for both turbine and electrical rotor shafts.

The diameter of the shaft is usually such that the critical speed does not come within 30 percents of the running speed. The critical speed of a shaft is that speed at which a very small eccentric mass will cause the shaft to deflect to a very great extent. This speed coincides with the natural frequency of vibration of the shaft. Hollow construction results in rotors of relatively light weight and high critical speed.

Blades of the turbine are constructed to withstand vibration, high temperature, shocks, and to resist deterioration by pitting, blunting, corrosion and crystallization. There are many types of blade materials, some of which are phosphor bronze, carbon steel, nickel steel, stainless steel, stainless iron and mild steel. Blades may be made by machining from solid bar, be drop forged, extruded or rolled; rolling is considered the best method. The diameters of the blade ring fixes are available for steam flow. The permissible stresses in the wheel and blade in turn fix the tip diameter. To reduce centrifugal stress in the blades, they are tapered, so that the cr4oss sectional area decreases steadily from the root to the tip. The greater the degree of taper the more can the stress be reduced. The use of a hollow blade at the exhaust end has made possible an increase in output of about 20 percent. The blades are also given a twist along its length. This permits smooth entry of steam at all points. The outer ends of the blades are shrouded with bands of suitable metal firmly secured by a projection on each blade, which passes through the band and is riveted over on the outside. This increases the rigidity of the balding.

7.8. Turbine Governing System

Turbine governing systems function to maintain one or more of the following functions:

1. Maintain turbine shaft speed constant
2. To adjust steam flow to match the load requirements or the turbo –generator
3. To maintain steam inlet pressure constant.

Whenever shaft load decreases with constant throttle valve opening, the turbine speeds up because it is getting more steam than what is required. Shaft speed can be maintained at same value by throttling steam flow to match the new load requirements. Similarly on shaft load increase. Its speed decreases and governor makes these changes automatically. In smaller turbines, the speed is usually sensed by a flyweight assembly consisting of two flyweights mounted on a plate which turns about a vertical axis driven by the turbine shaft through a worm and gear. The centrifugal force tilts the weights outwards, thereby compressing the stationary speeder spring and lifting up the speeder rod. The speeder rod movement is further hooked up to open and close the turbine throttle valve. The force exerted on speeder rod is dependents on centrifugal force of flyweights, which is equal to MRN^2

where

M- mass of flyweight

R- radius of their motion

N- rpm

If it be assumed that there is no speeder spring, then for one constant speed the centrifugal force(F) of flyweight will depend on radius R are accordingly the speeder rod or governor travel will also changes. R be plotted against governor travel for various speeds from 98 to 120%, we get the family of straight line curves as show in Fig 7.9. These curves actually converge at a common point if produced further on to left side. Slope of each speed curve has a scale reading in kg force/mm of governor travel. Selecting of spring rate is an important criterion for governor which will be obvious from below. If the spring is selected having its radius same as corresponding to 100% speed characteristic then control which be fine if the load is kept constant. If load increases, the shaft speed would reduce and weight develop less centrifugal force then the spring over the range of governor travel, weights would more than maximum in position thereby the steam value fully. As steam flow increases, sped would increase and slightly above 100% speed the spring's mechanical force would overbalance the weight's centrifugal force weights would fly to out limit, completely closing the steam valve. This would thus lead to hunting.

To avoid this situation the spring force must grow faster than the weight force as speed increases. i.e. spring rate should be steeper in flyweight scale as shown in fig 7.9. Assuming that the turbine to be running at 100% speed corresponding to point A, when load demand increases, the shaft speed decreases. The decreases force lets the balancing spring force lower the speeder rod to open the throttle valve wider, thereby increasing steam flow. The diminishing spring force matches weight force at a lower speed, but the governor will not travel to maximum in position because the weight exert more force than the spring below the new balance point. Similarly on load decrease, the governor cannot travel to completely out position as spring exerts greater force that the weights beyond the new point of balance.

It will be noted now that for each point in governor travel the turbine runs at a definite speed, being slower at full load and faster at no load. In fig 7.9, for the spring rate shown, speed varies by 4% from no load to full load. This value is known as speed droop or governor's regulation. It will be shown little later, how the speed can be kept constant at all loads by addition of a speed charger spring. It should be noted that the friction in the linkage etc. will not allow the governor to respond immediately and accordingly the governor does not behave as per actual spring rate line but on two parallel dotted lines depending on whether load is increasing or decreasing. The vertical distance between the upper and lower dotted lines in fig 7.9 is known as dead band or sensitivity of the governor i.e. Speed change necessary to produce a corrective movement in governor travel.

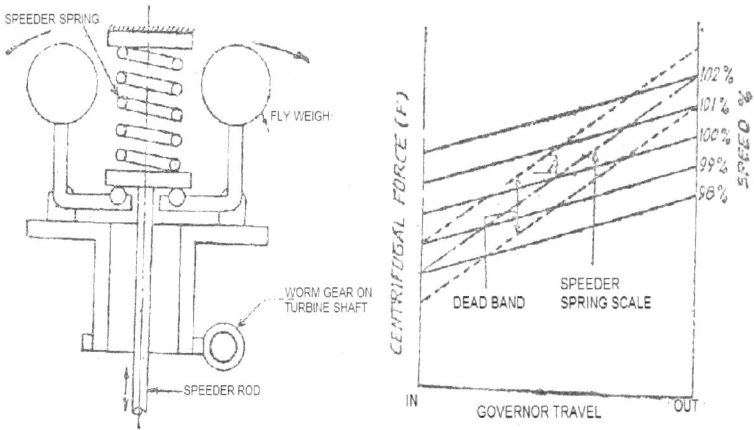

Fig. 7.9: Turbine Governing System

7.9. Method of Governing

Various method of speed governing of turbines is described below:

7.9.1. Direct Regulation

In case of very small turbines of 50 Kw rating with light regulation valves, it is possible to adoption direct sped regulation. In which the movements of governor is directly connected to turbine throttle valve by means of a lever AB having its fulcrum at O. when load decreases, governor speed increases and point A moves up and B down thereby closing the throttle valve to maintain desired speed. This is the simplest arrangement suitable only for very small turbine. For medium and large capacities the force required for operating then regulating valves is high and cannot provided directly by the governor centrifugal force of the weights. In such turbine, servomotors capable of exerting the required force for operation of large regulating valves is must.

7.9.2. *Indirect System of Governing*

It uses servomotor of the piston type. Under equilibrium condition i.e. constant speed operation, the control piston Pc of the pilot valve occupies a position midway between its travel, in which both the inlet and outlet valves of the pilot valve connecting it to servomotor are closed and throttle valve for these conditions occupies a certain fixed position.

Whenever load changes, governor sleeve changes its position causing displacements of the control piston Pc. As a result the oil from the oil pump enters either chamber C_1 or C_2 of the servomotor. If load decreases, oil enters the upper half portion i.e. chamber C_1 and throttle valve starts closing, reducing the quantity of steam through turbine. At the same time oil from chamber C_2 starts flowing out through the pilot valve port to drain. As the throttle valve closes, the point B of the lever AB starts moving down, lever AB operating with respect to point A as fulcrum and in the process displaces the control piston Pc downwards along with it. As soon as Pc occupies its original central position, entry of o8l to chamber C_1 is stopped and piston P_1 of servomotor again establishes new position of throttle valve according to load requirement. It will be appreciated that piston Pc does not required large amount of force. The force required to operate throttling valve is dependent upon size of piston P_1 and oil pressure. Lever AB is known as differential lever since with its help the piston Pc can be brought back to its central position. It will be appreciated that in this method speed at various loads does not remain exactly constant and is more at lower loads and less at higher loads. Of course, variation is very small.

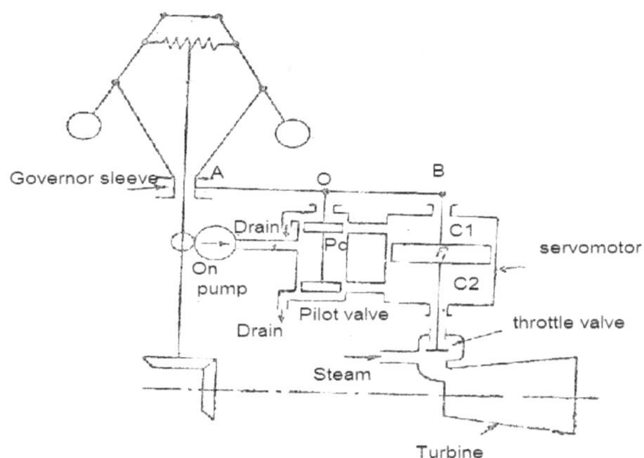

Fig. 7.10: Direct Regulation Governing

Fig 7.11 shows the variation of shaft speed at various loads. Curve S_1 shows the adjustment made to keep shaft speed at 100% corresponding to 100% load. For these settings, as the turbine load decreases, the shaft speed would increases and assuming governor's regulation or sped droop as 4% the speed at zero loads will be 104%. If a speeder changer spring be incorporated acting on the lever AB and if it is possible to change the spring forces on the lever by hand or remote controlled motor then it is possible to adjust it such that 100% speed is obtained at all the loads. In fig 7.11. Curves S_1, S2, S3, S4 shows the speed adjusted at 100% corresponding to 100%, 75%, 50%, and 25% load respectively by adjusting force of speed changer spring; if the force on this spring is constant corresponding to S_1, then at lower loads speed increases. In order to reduce it to 100% value, the speed changer spring should be loosened, allowing the steam valve to be closed a little more to hole sped at constant value of 100%

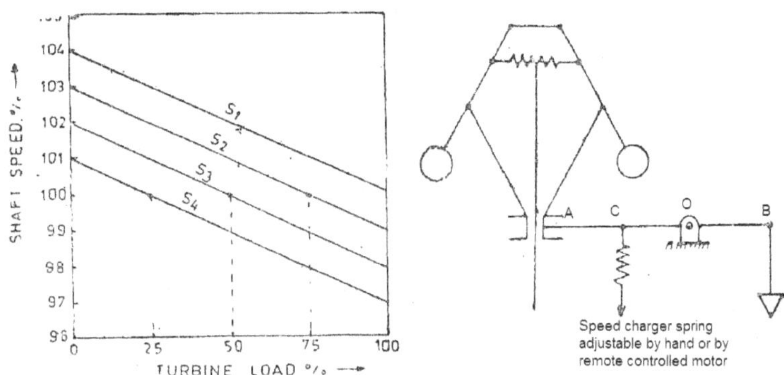

Fig. 7.11: Variation of Shaft Speed at Various Loads

7.10. Parallel Operation of Two Turbines

The sharing of load between two turbines feeding power into common grid depends directly on their governor droops. It is important to understand that paralleled synchronous generators run at the same speed as if they were connected mechanically. Whenever load requirement of grid changes, the speed will fall until total output of all units meets the new demand. Fig 7.12. Shows the droop characteristics of two turbines. Corresponding to 100% speed, these sets will take loads L_1 and L_2. If load increases and say sped falls to 99% then these sets will take up loads L_3 and L_4> The speed can be restored to 100% by adjusting the speed changers on both the units to lift their droop curves the same amount.

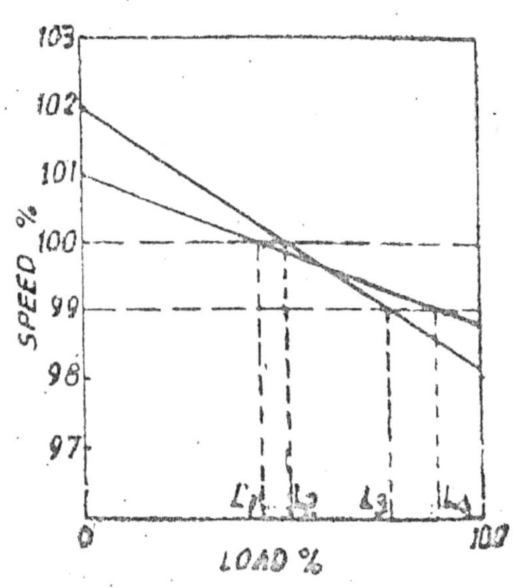

Fig. 7.12: The Droop Characteristics of Two Turbine

It is important to note that unit with least droop will always take largest share of the total load.

7.11. Hydraulic Governor

Many of the hydraulic governors utilize special centrifugal pumps which generate oil pressure proportional to square of shaft sped, instead of flyweights as speed sensing element. This primary oil pressure acts on diaphragm or "hydraulic speed governor" against the force of speed setting spring, which is compressed by sped changer. The speed changer is utilized to shift the droop curve. Arrangements are also made to limit travel of diaphragm by load limiting devices. The movement of diaphragm is transmitted by link mechanism to auxiliary follow-up pistons. The piston of auxiliary follow up piston is held in balance by is spring against oil pressure. The middle port of auxiliary cylinder at 'O' in Fig. 7.13. is feed from a trip circuit and is drained from a port formed between the piston and the sleeve. Depending upon port opening oil pressure gets stabilized corresponding to initial displacement of the piston and initial spring tension. The necessary auxiliary oil pressure provides a signal to hydraulic servomotor through its pilot. The piston of the hydraulic servomotor assumes a position corresponding to secondary auxiliary oil pressure and also actuates the sleeve of auxiliary follow up pistons, through the linkage mechanism. This fed back mechanism quickly stabilizes the position of the pilot valve and the piston of hydraulic servomotor.

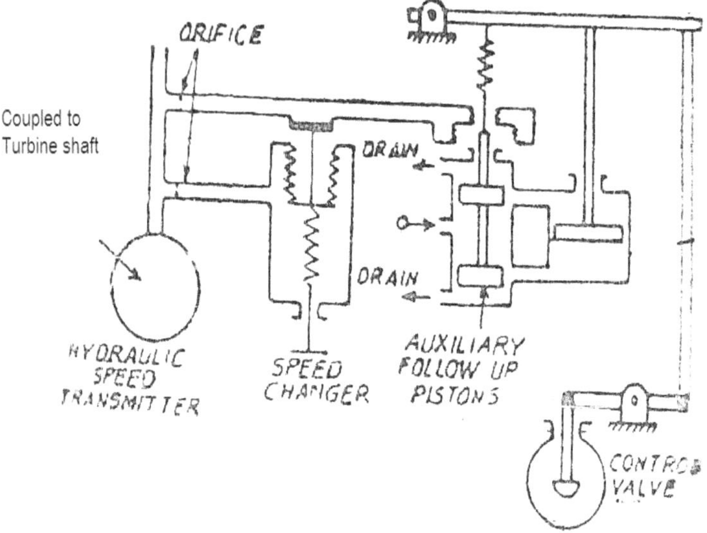

Fig. 7.13: Hydraulic Governor

The emergency stop valve is also operated by the governing oil and in case tripping is desired, the oil pressure is released thereby closing the valve. Over speed trip in the form of spring loaded unbalanced pin is also incorporated. When shaft speed exceeds the safe limit, centrifugal force snaps the pin to an outer position so it trips a lever that closes the emergency stop valve.

7.12. Simplified Operation of the Governor

The governor controls the fuel rack position through a combined action of the hydraulic piston and a set of mechanical flyweights, which are driven by the engine blower shaft. Figure 7.14 provides an illustration of a functional diagram of a mechanical-hydraulic governor. The position of the flyweights is determined by the speed of the engine. As the engine speeds up or down, the weights move in or out. The movement of the flyweights, due to a change in engine speed, moves a small piston (pilot valve) in the governor's hydraulic system. This motion adjusts flow of hydraulic fluid to a large hydraulic piston (servo-motor piston). The large hydraulic piston is linked to the fuel rack and its motion resets the fuel rack for increased/decreased fuel.

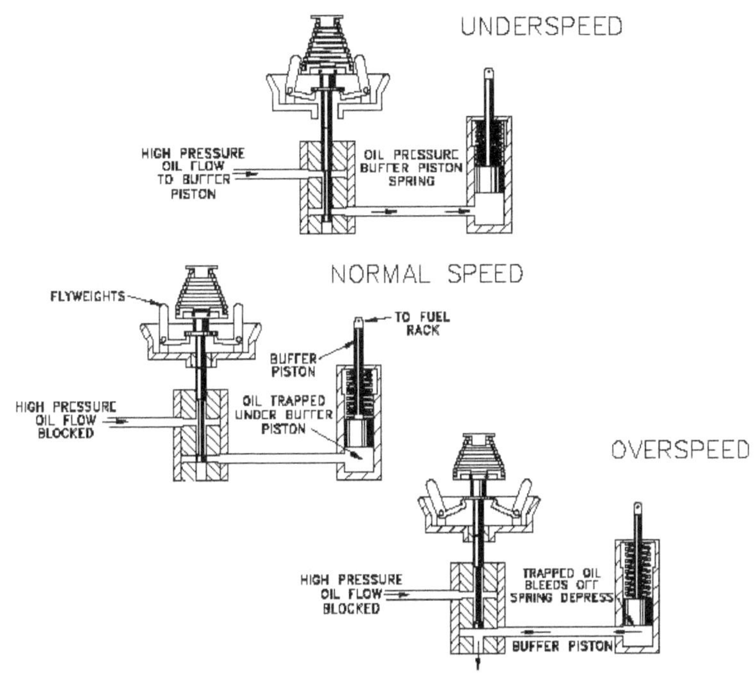

Fig. 7.14: Mechanical-Hydraulic Governor

7.13. Turbine Oil System

Mineral oil possessing the properties given below is recommended to be used as the fluid for actuating the governing system. The same oil is used for the purpose of lubrication. The oil –14 of Indian Oil Company or Mobile DTE medium generally meets the specification of the oil required for both the systems.

Specification of oil:

1. Specific gravity at 50°C   0.852
2. Kinematics viscosity at 50°C in centistokes   28
3. Neutralization number   0.2
4. Flash point in °C   201 (min)
5. Pour pint in ° C   -6.6(max)
6. Ash percentage by weight   0.01%
7. Mechanical impurities   Nil

8. Turbine Control System

8.1. Method of Governing

The object of governing is to maintain the speed of a turbine sensibly constant irrespective of load. The performance will depend, to a large extent, on the method used to regulate the steam supply to the turbine. The popular methods of governing are:

1. Throttle governing
2. Nozzle control governing
3. By-pass governing
2. Combination of 1 and 2
3. Combination of 1 and 3

8.1.1. Throttle Governing

Refer Fig. 8.1. In throttle governing, the steam pressure at which steam is admitted to the turbine is reduced at part loads. The mechanism of throttle governing is simple but thermodynamically it is not efficient due to available heat drop getting reduced in the irreversible throttle process. Thus it is used on small turbines. The steam flow to the turbine is throttled by balanced throttle valve actuated by a centrifugal governor. The effort of the governor may not be sufficient to move the valve against friction in bigger units. In that case an oil relay is incorporated, and a small force produced by governor for a small change of speed in magnified to actuate the throttle valve.

A simple differential relay is shown in Fig. 8.1. The throttle valve is moved by a relay piston. A floating differential lever is fixed to its one end and at some intermediate point on the floating differential lever, a piston valve is fixed. The pilot piston valve consists of two piston valves covering ports without any overlap. The piston valves are also operated by lubricating oil supplied by a pump at about 3 to 4 bar. The oil from this chamber is returned to the oil drain.

Operation

Let the turbine work at full rated load. The throttle valve will open to take care of this load, at rated constant speed. If the load is reduced, the energy supplied to the turbine will be in excess and the turbine rotor will accelerate. Thus the governor sleeve will lift. Since the throttle valve position is assumed same momentarily, the pilot piston valve spindle will get lifted opening the upper port to oil pressure and lower port to oil return. The relay piston will thus close the throttle valve partially. The lowering of throttle valve spindle will lower the pilot piston spindle and close the ports. As soon as the ports are closed, the relay piston gets stabilized in one position corresponding to the reduced load.

The change in available enthalpy for throttle governing is shown in Fig. 8.2.

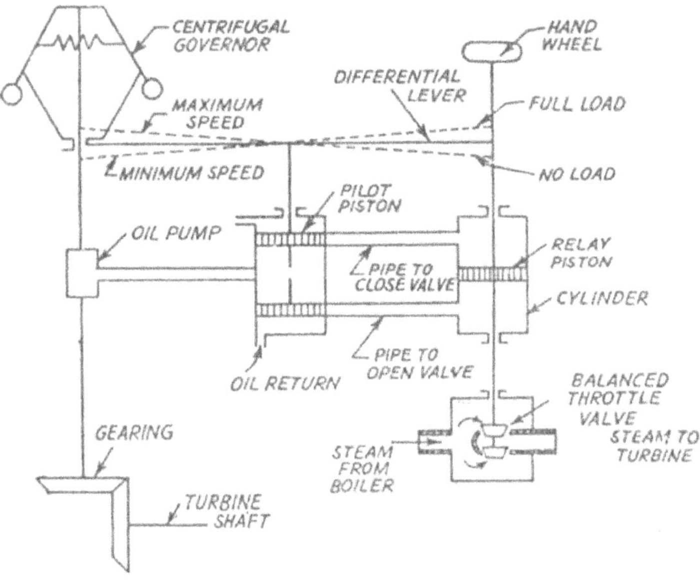

Fig. 8.1: Simple Differential Relay

8.1.2. Nozzle Control Governing

The principle of nozzle control is accomplished by uncovering as many steam passages as are necessary to meet the load by poppet valves. In applying the principle to land turbines which are almost entirely governed by automatic devices, various arrangements of valves and groups of nozzles are employed. Three arrangements are shown diagrammatically in Figs. 8.3, 8.4 and 8.5. An arrangement often used with large steam turbines and with turbines using high pressure steam is shown in Fig. 8.3. The nozzles are divided into groups. N_1, N_2 and N_3 under the control valves V_1, V_2 and V_3 respectively. The number of nozzle groups may vary from three to five or more. Fig. 8.4 differs from Fig. 8.3 only in that the nozzle control valves are arranged in a casting forming part of the cylinder or bolted thereto and containing passages leading to the individual nozzle groups. Although this arrangement is compact, the nozzles are confined to the upper half of the cylinder and the arc of admission is limited to 180° or less. The number of nozzle groups varies from four to twelve. Fig. 8.5 shows an arrangement sometimes employed. For the sake of illustration only four groups of nozzles are shown. The group of nozzles N_1 is under the control the valve V_1 through which all the steam entering the turbine passes.

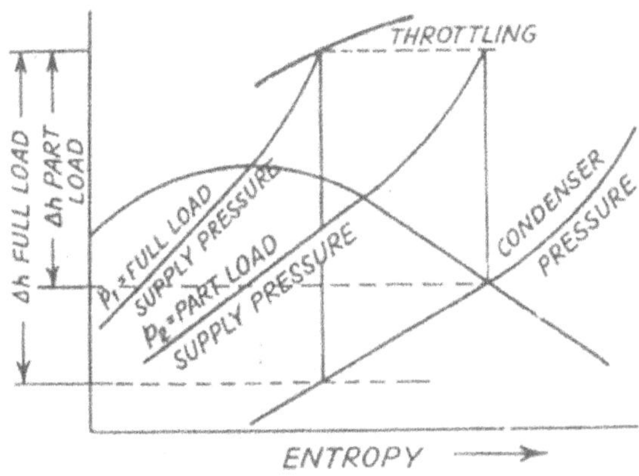

Fig. 8.2: Change in Available Enthalpy for Throttle Governing

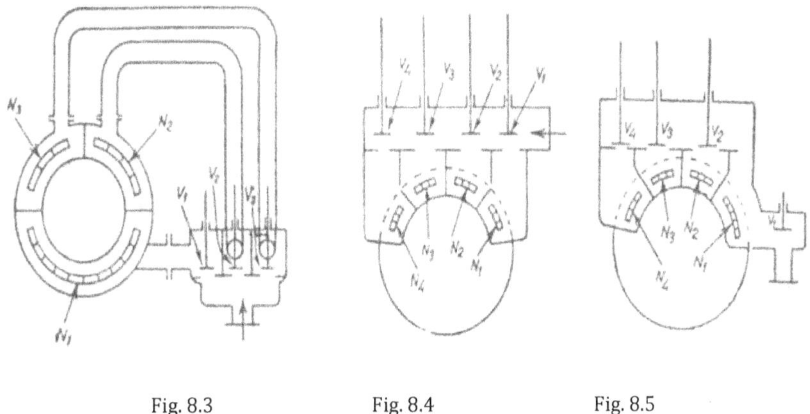

Fig. 8.3   Fig. 8.4   Fig. 8.5

Further admission of the steam is through the valves V_2, V_3 etc., in turn. In some instances, the nozzle group N_1 has been arranged in the lower half of the turbine and supplied with steam through a valve V_1 upto say ½, load. For loads greater than ½ load, a further supply of steam is admitted through the valves V_2 and V_3 etc.

8.1.3. By Passing Governing

In modern impulse turbine and specially those operating at very high pressures, the high pressure turbine comprises a number of stages of comparatively small mean diameter. All such turbines are usually designed for a definite load termed as economic load, at which the thermodynamic efficiency ratio of the turbine is maximum. In case of power station turbines, the economic load is made about 80% of the maximum continuous rating (M.C.R.), which is maximum output, the generator is capable of giving continuously.

Owing to the very small enthalpy change in the first stage it is not possible to employ nozzle control governing efficiently. Furthermore, it is desirable to have full admission in the high pressure stages at the economic load so as to eliminate the partial admission losses. Hence it is not possible to admit through additional nozzles in the first-stage the 15 to 17 percent extra steam necessary to generate the full output. In fact in most cases it would be quite impossible to do so because the steam pressure in the first stage wheel chamber would increase so much, due to the restriction, at the second-stage nozzles, that it would require to be greater than the initial pressure.

The difficulties of regulation are overcome by employing by-pass governing. The principle of by-pass governing is shown diagrammatically in Fig. 8.6. All the steam entering the turbine passes through the main throttle valve which is under the control of speed governor and enters the nozzle box or the steam chest. In certain cases, this would suffice for all loads upto economical load, the governing being effected by throttling. For loads greater than the economic load, a by-pass valve is opened allowing steam to pass from the first stage, nozzle box into the steam belt and so into the nozzles of the fourth stage. The by-pass valve is not opened until the lift of the throttle valve exceeds a certain amount; also as the load diminishes the by pass valve closes first. The by-pass valve is under the control of speed governor for all loads within this range.

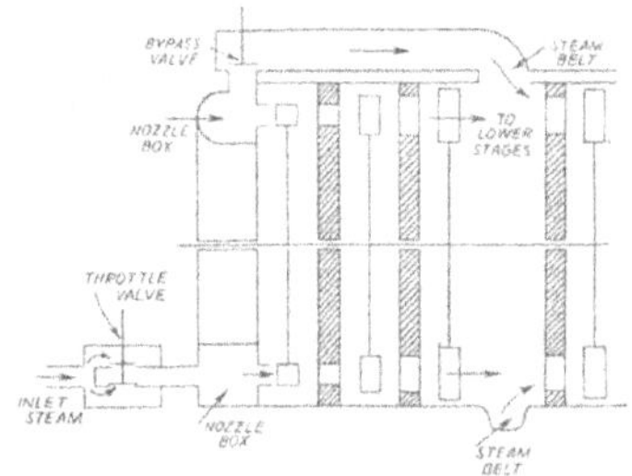

Fig. 8.6: By-Pass Governing

The number of by-pass valves is not always restricted to one. The outline of a 100 MW steam turbine is shown in Fig.8.7. There are three-cylinders arranged in tandem and containing the following number of stages.

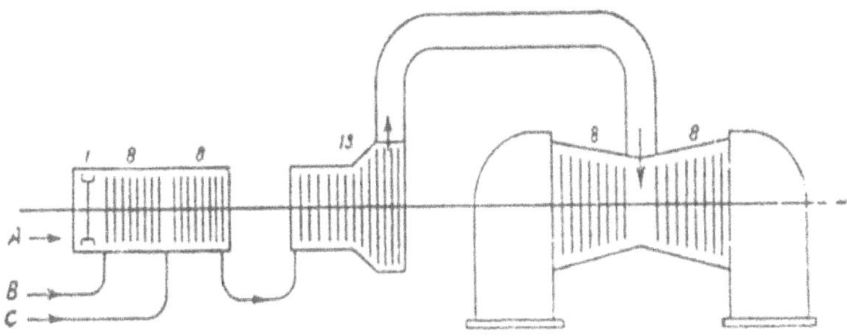

Fig. 8.7: 100 Mw Steam Turbine

High pressure - 1 (2 row wheel) + 16 (single row) stages

Intermediate pressure: 13 stages.

Low pressure: a double flow turbine each half with 8 single row wheels

For loads upto 60% of maximum continuous rating, steam is admitted before the velocity stage, for loads upto 80% steam is admitted at B after the velocity stage, while for the full output of the turbine steam is admitted at C, the first nine stages being by-passed.

8.1.4. Back Pressure and Pass-out Turbines

There are several industries, such as paper making, textile, chemical, dyeing; sugar refining etc. in which there is dual demand for power and steam for heating and process work. It was formerly the practice to generate steam for power purposes at a moderate pressure and to generate saturated steam for process work at a pressure which gave the desired heating temperature. Such a system is obviously wasteful for the total quantity of heat supplied to the steam which is used for power purposes, quite 70% will normally be carried away by the cooling water. Admittedly, if the engine or turbine is operated with normal exhaust pressure, 'then the temperature of the exhaust steam is too low to be of any use for heating purposes, but by suitable modification of the initial pressure and the exhaust pressure it would be possible to generate the required power and still have available for process work a large quantity of heat in the exhaust steam.

8.1.5. Back Pressure Turbines

The back pressure turbine may be used in cases where the power, which may be generated by expanding steam from an economical initial pressure down to the heating pressure, is equal to, or greater than the power requirements. The layout of such a plant is shown diagrammatically in Fig. 8.8. Steam is generated in the boiler at a suitable working pressure and admitted to the turbine. file exhaust steam from the turbine will normally be super-heated and in most cases is not suitable for process work partly because it is impossible to control its temperature, since it will vary with initial superheat and partly because of the fact that the rate of heat transfer from superheated steam to the heating surfaces is lower than that of saturated steam. For these reasons a desuper-heater is often used. A jet of water, thermostatically controlled, is sprayed into the entering steam; the steam is cooled and spray water evaporated. The new saturated steam enters the heaters and is entirely condensed.

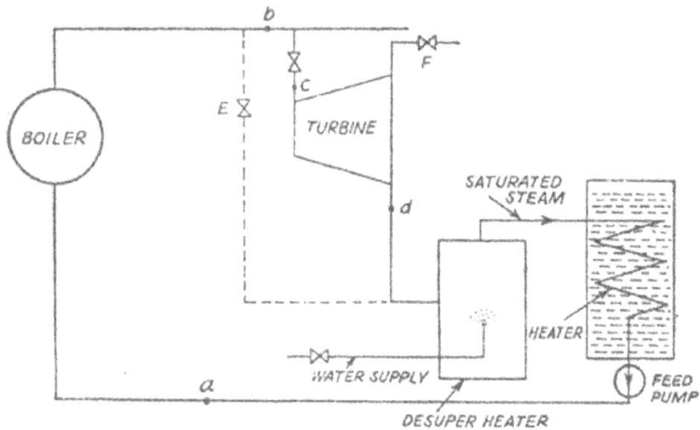

Fig. 8.8: Layout of Back Pressure Turbine Plant

The condensed steam may or may not be returned to the boilers depending upon the conditions. It is unlikely that the steam required for power generation will always be equal to that required for process work and some means for controlling the exhaust steam pressure must be employed if variations in pressure and, therefore, of the steam saturation temperature are to be avoided.

The method of control employed depends upon circumstances. If the back pressure turbine is only a power unit, then it is fitted with an ordinary centrifugal governor and the quantity of available exhaust steam is then controlled by the load on the turbine. If the available exhaust steam is too small, live steam may be passed through the reducing valve E into the desuperheater. If the quantity of exhaust steam is in excess of requirements, then the excess steam may be blown to atmosphere, or into feed tank etc. through the valve F.

Valves E and F may be operated by hand to maintain the exhaust pressure constant or as is more usual, by automatic control gear. If however the back pressure turbine is operating in parallel with other machines, then its output is controlled entirely by the heat load. In addition to the normal speed governor, there is a pressure regulator which controls the supply of steam to the turbine so as to maintain a suitable constant exhaust pressure.

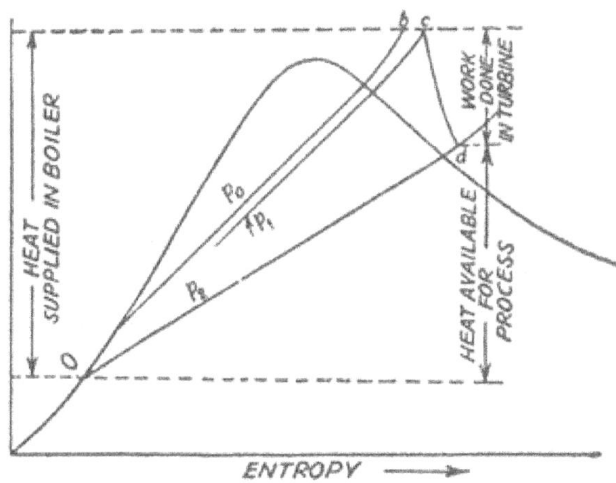

Fig. 8.9: The Thermodynamics of Back Pressure Turbine

The thermodynamics of back pressure turbine is extremely simple and may be explained with reference to Fig. 8.9 which is labeled to correspond Fig. 8.8 so far as the state of working substance is concerned. The boiler pressure is P_0, the pressure at inlet to the turbine is P_1 and the heating steam pressure is P_2 be shows the throttling at throttle valve cd the expansion in the turbine and do the subsequent condensation in the process heaters.

8.1.6. Pass Out Turbine

In many cases, the power available from a back pressure turbine through which the whole of the heating steam flows is appreciably less than that required in the factory. This may be due to small heating or process requirements, to a relatively high back pressure or combination of both. In such a case it would be possible to install a back pressure turbine to generate the extra power; but it is possible and usual to combine the functions of both machines in a single turbine with obvious advantages and economics. Such a turbine is shown diagrammatically in Fig. 8.10. Live steam enters the turbine and expands through the high pressure stages before the extraction branch. Here a certain quantity of steam is continuously being extracted for heating purposes, the remainder passing through the pressure control valve into the low pressure part of the turbine.

The control gear must be designed to meet the following requirements:

1. The speed of the turbine must be kept constant.
2. The pressure of the heating steam must be kept sensibly constant, both of these regardless of variations in power and heating loads.

These requirements are adequately met by the use of (a) a speed governor of normal centrifugal type which controls. the admission of the high pressure steam to the turbine, and (b) a pressure regular responsive to change of pressure of the extracted heating steam and controlling the admission of the steam to the low pressure stages. A change of either power load or heating load will cause both governors to operate. For example, suppose the generator load is increased, this causes a certain temporary drop of speed. In response to this the speed governor increases the lift of the high pressure throttle valve, thus allowing more steam to enter the turbine. If no change were made in the position of the pressure control valve the pressure in the heating steam pipe would now rise. This pressure rise actuates the oil relay in such a way as to increase the lift of the pressure control valve and allow more steam to flow through the low pressure to the condenser. For a constant heat load, the mass of heating steam will remain nearly constant, it will vary slightly due to variations in enthalpy: therefore, the increase in steam flow through the low pressure stages will be practically equal to that through the high pressure stages. Thus, with constant heat load, an increase in power load is followed by an increase in lift of the throttle and pressure control valves. Conversely, a reduction in power is followed by reduced lift of both valves.

Suppose now that there is sudden increase in heat load due, say to coupling up another heater unit. The heating steam pressure will fall and so cause the lift of the pressure control valve to be reduced. This reduces the flow of steam through low pressure stages with the result that the less power is developed in those stages, and as the power of the high pressure stages is yet hardly affected, the speed of the turbine begins to fall. This causes high pressure throttle to open, so admitting greater quantity of steam to the turbine as a whole. It will be seen that a change in the heating load causes the lifts of the main throttle and pressure control valves to vary in opposite ways.

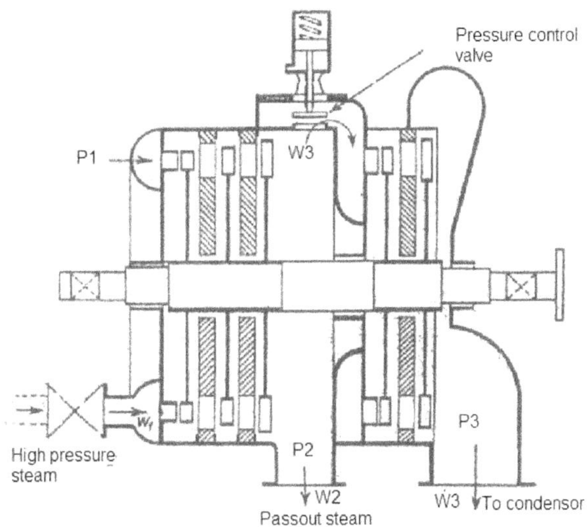

Fig. 8.10: Pass-Out Turbine

When the speed and the pressure governor operates independently of each other, it will be apparent that, provided the driven alternator is not connected electrically to any system, an increase in power load brings about an increase in pass–out steam pressure, whereas an speed. There is also a tendency to hunting.

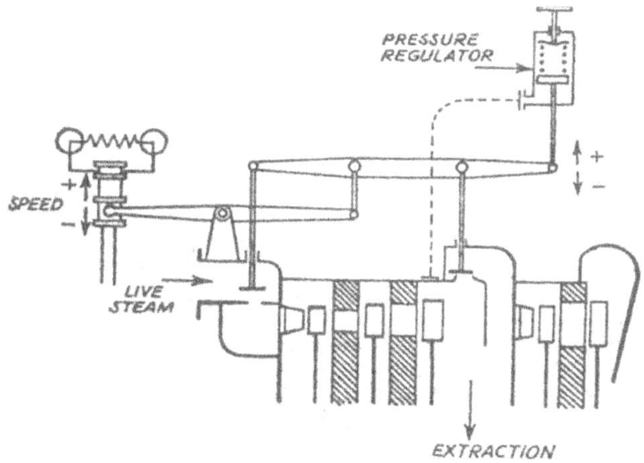

Fig. 8.11: Inter Connecting the Governor Gear and Pressure Regulating Gear Arrangement

This may be overcome by interconnecting the governor gear and the pressure regulating gear in such a way that both the throttle valve and the pressure control valve are operated simultaneously and in correct direction. One form of such gear is shown diagrammatically in figure 8.11. In order to simplify the diagram, the oil relays have been omitted. It will be seen that a drop in speed causes both valves to lift whereas an increase in heat load in producing a fall in heating steam pressure causes the pressure control valve partially to close and the throttle to open further.

8.2. Programmable Logic Controller

It can be defined as a digital electronic device that uses a programmable memory to store instructions and to implement functions such as logic, sequencing, timing, counting and arithmetic in order to control machines and processes Fig. 8.12. The term logic is used because the programming is primarily concerned with implementing logic and switching operations. Inputs device S, e.g. switches, and output devices, e.g. motors, being controlled are connected to the PLC and then the controller monitors the inputs and outputs according to this program stored in the PLC by the operator and so controls the machine or process. Originally they were designed as a replacement for hard-wired relay and timer logic control systems. PLCs have the great advantage that it is possible to modify a control system without having to rewire the connections to the input and output devices, the only requirement being that an operator has to key in a different set of instructions. The result is a flexible system which can be used to control systems which vary quite widely in their nature and complexity.

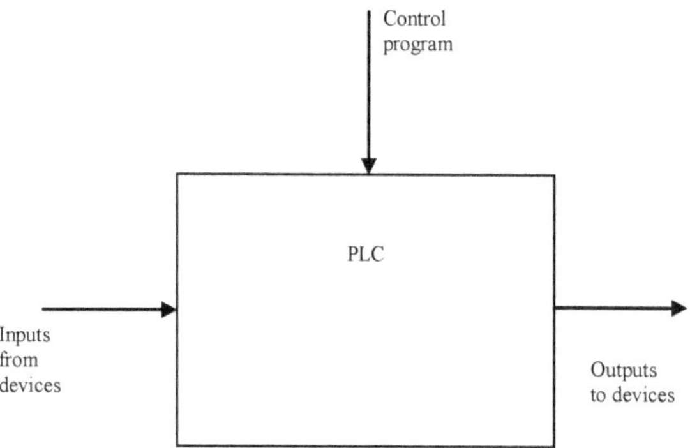

Fig. 8.12: Programmable Logic Controller

PLCs are similar to computers but have certain features which are specific to their use as controllers.

These are:

1. They are rugged and designed to withstand vibrations, temperature, humidity and noise.
2. The interfacing for inputs and outputs is inside the controller.
3. They are easily programmed and have an easily understood programming language. Programming is primarily concerned with logic and switching operations.

PLCs were first conceived in 1968. They are now widely used and extend from small self-contained units for use with perhaps 20 digital input/outputs to modular systems which can be used for large numbers of inputs/outputs, handle digital or analogue inputs/outputs, and also carry out PID control modes.

8.2.1. Basic Structure

Figure 8.13 shows the basic internal structure of a PLC. It consists essentially of a central processing unit (CPU), memory, and input/output circuitry. The CPU controls and processes all the operations within the PLC. It is supplied with a clock with a frequency of typically between 1 and 8 MHz. This frequency determines the operating speed of the PLC and provides the timing and synchronization for all elements in the system. A bus system carries information and data to and from the CPU, memory and input/output units. There are several memory elements: a system ROM to give permanent storage for the operating system and fixed data, RAM for the user's program, and temporary buffer stores for the input/output channels.

The programs in RAM can be changed by the user. However, to prevent the loss of these programs when the power supply is switched off, a battery is likely to be used in the PLC to maintain the RAM contents for a period of time. After a program has been developed in RAM it may be loaded into an EPROM memory chip and so made permanent. Specifications for small PLCs often specify the program memory size in terms of the number of program steps that can be stored. A program step is an instruction for some event to occur. A program task might consist of a number of steps and could be, for example: examine the state of switch A, examine the state of switch B, if A and B are closed then energies solenoid P which then might result in the operation of some actuator. When this happens another task might then be started. Typically the number of steps that can be handled by a small PLC is of the order of 300 to 1000, which is generally more than adequate for most control situations.

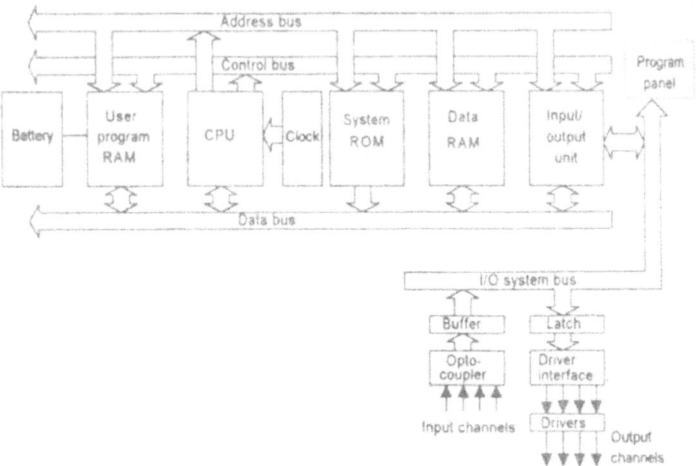

Fig. 8.13: Architecture of a PLC

The input/output unit provides the interface between the system and the outside world. Programs are entered into the input/output unit from a panel which can vary from small keyboards with liquid crystal displays to those using a visual display unit (VDU) with keyboard and screen display. Alternatively the programs can be entered into the system by means of a link to a personal computer (PC) which is loaded with an appropriate software package. The input/output channels provide signal conditioning and isolation functions so that sensors and actuators can be generally directly connected to them without the need for other circuitry. Figure 8.14 shows the basic form of an input channel. Common input voltages are 5 V and 24 V.

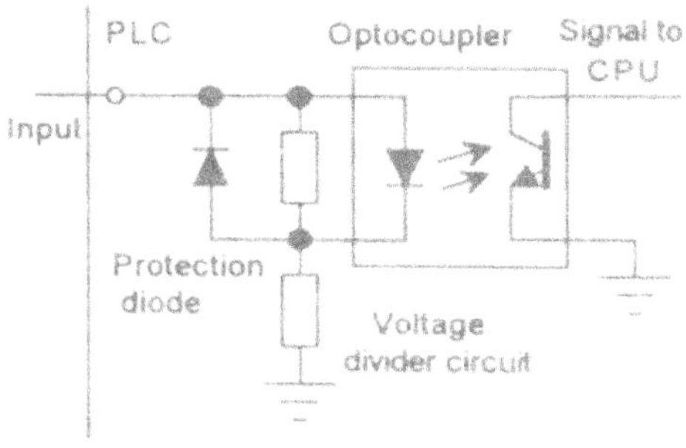

Fig. 8.14: Input Channel

Common output voltages are 24 V and 240 V. Outputs are often specified as being of relay type, transistor type or trial type. With the relay type Fig. 8.15, the signal from the PLC output is used to operate a relay and so is able to switch currents of the order of a few amperes in an external circuit. The relay isolates the PLC from the external circuit and can be used for both DC and AC switching. Relays are, however, relatively slow to operate. The transistor type of output Fig. 8.16 uses a transistor to switch current through the external circuit. This gives a faster switching action. Opt isolators are used with transistor switches to provide isolation between the external circuit and the PLC. The transistor output is only for DC switching. Trial outputs can be used to control external loads which are connected to the AC power supply. Opt isolators are again used to provide isolation.

8.2.2. Input/output Processing

The basic form of programming commonly used with PLCs is ladder programming. This involves each program task being specified as though a rung of a ladder. Thus such a rung could specify that the state of switches A and B, the inputs, be examined and if A and B are both closed then a solenoid, the output, is energised.

The sequence followed by a PLC when carrying out a program can be summarised as:

1. Scan the inputs associated with one rung of the ladder program.
2. Solve the logic operation involving those inputs.
3. Set/reset the outputs for that rung.
4. Move on to the next rung and repeat operations 1, 2, 3.
5. Move on to the next rung and repeat operations 1, 2, 3.
6. Move on to the next rung and repeat operations 1,2,3.

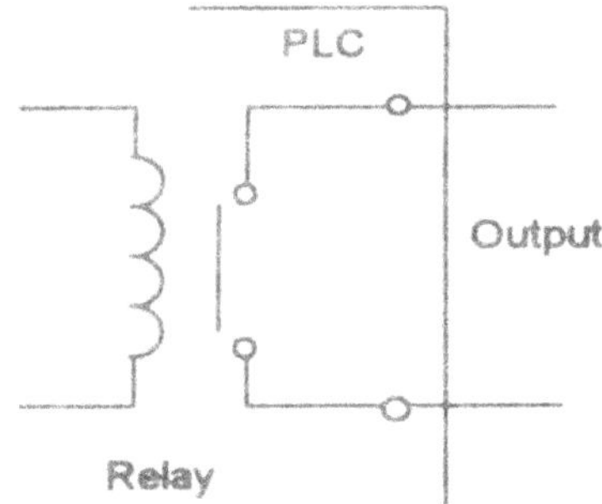

Fig. 8.15: Relay of Output

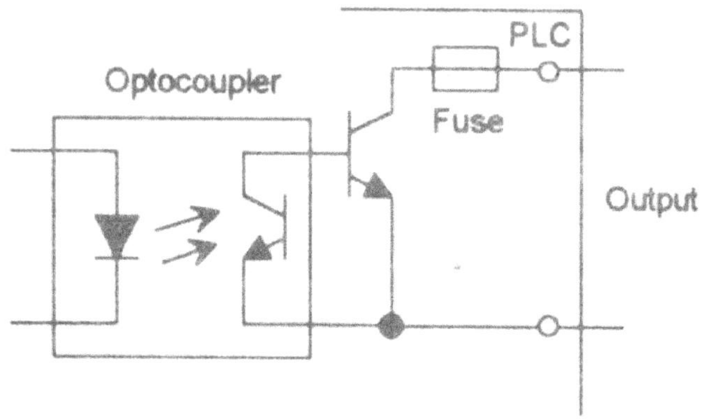

Fig. 8.16: Transistor Form of Output

And so on until the end of the program.

Each rung of the ladder program is thus scanned in turn.

There are two methods that can be used for input/output processing:

8.2.2.1. Continuous Updating

This involves the CPU scanning the input channels as they occur in the program instructions. Each input point is examined individually and its effect on the program determined. There will be a built-in delay, typically about 3 ms, when each input is examined in order to ensure that only valid input signals are read by the microprocessor. This delay enables the microprocessor to avoid counting an input signal twice, or more frequently, if there is contact bounce at a switch. A number of inputs may have to be scanned, each With a 3 ms delay, before the program has the instruction for a logic operation to be executed and an output to occur. The outputs are latched so that they retain their status until the next updating.

8.2.2.2. Mass Input/output Copying

Because, with continuous updating, there has to be a 3 ms delay on each input, the time taken to examine several hundred input/output points can become comparatively long. To allow a more rapid execution of a program, a specific area of RAM is used as a buffer store between the control logic and the input/output unit. Each input/output has an address in this memory. At the start of each program cycle the CPU scans all the inputs and copies their status into the input/output addresses in RAM. As the program is executed the stored input data is read, as required, from RAM and the logic operations carried out. The resulting output signals are stored in the reserved input/output section of RAM. At the end of each program cycle all the outputs are transferred from RAM to the output channels. The outputs are latched so that they retain their status until the next updating.'"

8.2.3. *Temperature Control System*

An on–off temperature control fig 8.17 in which the input goes from low to high when the temperature sensor reaches the set temperature. The output is then to go from on to off. The temperature sensor shown in the figure is a thermistor connected in a bridge arrangement with output to an operational amplifier connected as a comparator. The program shows the input as a normally closed pair of contacts, so giving the on signal and hence an output. When the contacts are opened to give the off signal then the output is switched off.

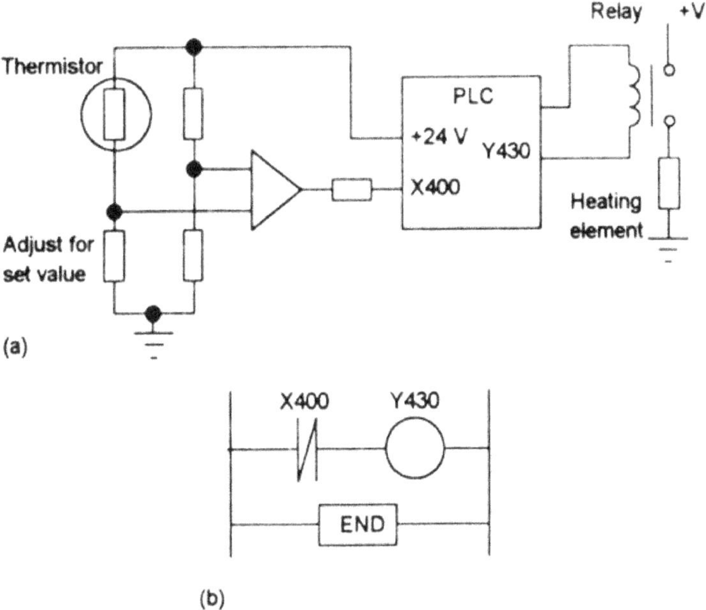

Fig. 8.17: Temperature Control System

Such ladder programs can be entered from special keypads or selected from a monitor screen by using a mouse. They can also be specified by using a mnemonic language. However they are entered, the programs are then translated by the PLC into machine language for the benefit of the microprocessor and its associated elements.

9. STEAM TURBINE COMMISSIONING, START UP AND SHUT DOWN PROCEDURE

9.1. Steam Turbine Commissioning

During the final stages of erection of turbines all internal accessible parts are freed of foreign matter like rust, scale, dust etc. by brushing, or blowing steam, compress3ed air or water etc. The whole of lub oil and governing oil system is thoroughly cleaned and closed. The feed heating and condensate system is first flushed out with water and then the stem side is alkaline cleaned by circulating caustic soda (200ppm) and trisodium phosphate (100 ppm) sin water at 99°c or may be soaked for four hours and then flushed out with hot dematerialized water. The water side   (excluding pumps) is cleaned by using 3% solution of citric acid in D.M. water at about 95°C by admitting steam into the deaerator. As soon as electrical and steam supplied are on, all auxiliaries should be tried out. Vacuum is then raised after sealing shaft glands and the system checked for leaks, and ejector operation also checked. Proper lagging of H.P. piping and turbine cylinders should be ensure to prevent distortion due to irregular thermal stress.

9.2. Turbine Installation and Operation

Basically the layout and installation of a steam turbine unit consist of work connected with

1. Foundation
2. Steam connection from high-pressure header
3. Exhaust connection
4. Extraction steam piping
5. Auxiliary piping (drip and drain, gland, oil)
6. Leads to remote instruments and controls
7. Electrical circuits for the generator and exciter

Small turbine and their driven equipment are customarily mounted on a rigid cast or welded bedplate which itself must be properly supported by a foundation of the power plant engineer's design. This can be a solid concrete block type for noncondensing unit, as well as for small condensing turbines with side exhaust. Large turbines have fore and aft support are in the form of brackets or pedestals and are carried on prepared foundations designed against specified allowable deflection under load. Condensing turbines usually exhaust downward and therefore require on open form of foundation, for space must be available immediately below the turbo-alternator for such items as condenser, extraction steam piping, generator cooler, conduits for generator electrical loads, etc.

The high–pressure steam lead must be installed with attention to the need of adequate, yet flexible support, thermal expansion, and the drain of condensate, the latter especially when warming up the pipe system preparatory to a start. The most damaging thing that can happen to a turbine is to receive a slug of water along with the steam when the blades are rapidly moving. A steam separator should be located in the supply line close to the turbine and any inverted expansion bends receives special drainage during warm-ups. Auxiliary drive turbines of less than 25hp may dispense with the separator if their lead is taken from the top of the header and if an upstream drain line is provided at the throttle valve. The configuration of piping necessary to connect the auxiliary to a source of steam, together with the small size of such piping, is usually adequate to absorb thermal expansion without imposing a stress on the turbine to which it is connected. But the large turbine with its short steam lead, large and thick-walled ought to be installed with spe4cial long-radius bends to absorb expansion. Typical steam leads from headers above or below the turbine room floor are shown in fig 9.1.

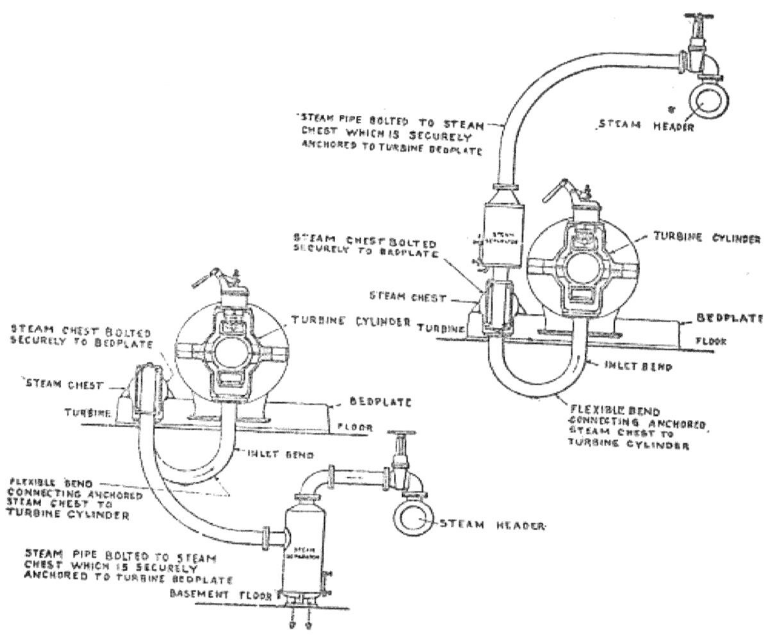

Fig. 9.1: Example of Main Steam Lead to Large Turbine

The bleeder piping is of relatively small size, usually having sufficient bends between turbine and heater to absorb expansion. There may be no exhaust piping as such, for the condenser inlet flange is frequently bolted directly to the turbine exhaust flange. Sometimes a flexible expansion joint is inserted at that point all depending on the method used to support the condenser weight. Small piping, some of which may be supplied by the turbine manufacturer, will consist of:

1. Condensate drains from casing, steam chest, etc.
2. Oil piping for lubrication and governing system.
3. Gland piping for leak-off steam of high –pressure gland and for supplying water or low –pressure steam to exhaust –end gland.
4. Pressure lines to instruments for steam and oil pressures.

9.3. Operating Supervision

Smaller turbines will have fewer of these, the ultimate in simplicity being the small, mechanical drive turbine with throttle valve, casing and throttle drain, and steam supply pressure gauge.

The starting of a noncondensing turbine is done by slowly opening the throttle valve until sufficient steam is striking the blades to overcome static friction. Then the shaft starts to rotate and the throttle should be adjusted so that a slow speed(200-500) will be maintained until pipe lines, casing, and rotor are thoroughly warmed up. But even before this is done the steam line should be thoroughly drained of condensate, the oil supply checked in the reservoir, and the driven machine inspected for "ready" condition. The driven with an oil relay in the governing system, and no auxiliary oil pump that can be previously started, the operator will need to hold the governor valve open against its spring force until the shaft- driven oil pump can produce sufficient oil pressure on the operating piston to take over. When the machine is warm and drains have been closed off, if oil pressure gauges show normal indication the throttle may be slowly opened until then speed governor takes over, then completely opened. At this point it is desirable to test the over speed trip, even though it may mean repeating the starting cycle. Two forms of trip test are:

1. Operate the trip trigger manually, which verifies the valve mechanism but not the centrifugal actuator
2. Overpower the governor, or with speed changer raise the shaft speed until the over speed element operates the trip release. This tests the entire trip system.

To stop a noncondensing turbine one merely needs to close the throttle valve.

9.4. Basic Turbine Starting Procedure

a.   Start auxiliary oil pump and check oil pressure

b.   Check level in oil reservoir

c.   open gland leak off valves

d.   Open all cylinder drain valves

e.   Drain condensate from the main steam header and the steam leads

f.   Establish circulating- water flow through condenser

g.   start condensate pump

h.   Establish seal on HP gland for starting condition

i.   Establish seal on LP gland for starting condition

j.   Start condenser air ejector and close vacuum breaker

k.   Close cylinder drains to stages under vacuum

l.   With partial vacuum established quickly admit enough steam to start rotor and then shut off

m.   Listen for rubs on casing and at seal locations

n.   If no rubs are evident, admit enough steam to establish rotor speed of about 200rpm. Maintain about one-half hour to warm up rotor and casing evenly.

o.   Trip emergency hand control to check operation

p.   Reestablish steam flow and slowly increase speed towards rated rpm during next 15 minutes. If rotor vibrates severely decrease speed and continue warming up until no objectionable vibration appears on speed increase

q.   Adjust HP and LP seals for operating condition

r.   When cylinder condensation ceases close drain valves

s.   Turn on cooling water to oil cooler to maintain about 43ºC outlet oil temperature

t.   AS turbine reaches rated speed make sure that the governor takes control

u.   Place unit on line quickly and apply about 20%load

v.   Open bleed-line valves and place heaters in operation

9.5. Turbine Shutdown Procedure

a.   Reduce turbine load gradually to zero and quickly take the unit off the line

b.   Close bleed-line valves and take heaters out of service

c.   Shut off steam by manual tripping of over speed trip

d.   Open vacuum breaker

e.   Shut off air ejector

f.   Check that auxiliary oil pump starts at proper speed

g.   shut off gland seal water

h.   shut down condensate pump

i.   shut off gland sealing steam

j.   open all atmospheric drains

k.   shut off water to oil coolers

l.   shut down condenser circulating- water pumps

m.   Keep auxiliary oil pump in operation until unit is cool

9.6. Starting Up of Turbine

The procedure given below is a typical one to explain the principles involved. Since actual details vary from manufacturer to manufacturer, the manufacturer recommendation should be followed strictly.

After carrying out preliminary checks on turbine, and operation of interlock and protection schemes, the first step could be to ensure proper drainage of main steam piping and turbine to prevent thermal distortion etc. The next step could be to bring in oil system by starting oil pumps and starting the barring year. Then shaft glands are sealed and vacuum is built with the quick start sealing ejector. Steam supply should be the minimum required for efficient sealing as excess of it may cause shaft warping and vibration stop valve or as the steam is then admitted to turbine by opening specified, after ensuring that control on emergency valve bypass(governor) valves are set. Appropriately as per recommendation of the manufacturer. The acceleration of turbine rotor will soon take the machine above the barring gear speed when the barring gear gets disengaged automatically. The speed is then increased steadily and fairly gradually to about 300rpm to enable an oil film to build up on the bearing surface and the internals of the turbine are soaked for sufficient time.

The speed is then increased steadily to about 2/3 of full speed at a rate recommended by the manufacturer. The critical speed region should be passed very quickly. During the period of raising speed from 300 to 2000 rpm careful checking should be done for rough running, rubbing, squealing or other abnormal sounds. A close watch axial shaft, differential expansion, turbine expansion and other turbo supervisory instruments and the metal temperature indications to ensure steady warming up of turbine. The speed is further raised and brought to grid frequency and the units synchronize the unit. After that load is increased and governor comes into operation.

9.7. Starting Turbine from Cold Condition (Cold Start-up)

If the temperature of lower part of HPT casing at the time of starting isles that 150°C, then it is called cold start of turbine. During starting of turbine it is very essential to keep a close watch on turbo supervisory parameters viz., eccentricity, vibrations, metal temperatures cylinder expansion and differential expansion. Before admitting steam to the turbine, the rotor of turbine is put on barring gear,lub oil temperature brought to 45±5ºC and a close watch kept on shaft eccentricity. I f eccentricity is more than 0.07mm, then rotor speed should not be increased. If during turbine running, the eccentricity exceeds 0.20mm, the turbine should be shut down immediately and put on barring gear. Turbine can be restarted only if the fault has been located and attended to. This holds for all types of faults. The barring gear gets disengaged automatically if turbine speed exceeds 3.4 rpm. If during the process of raising the speed of rotor, bearing vibrations approach 40 microns, the turbine should be shut down and the rotor should be put on the barring gear. It speed is to be decreased quickly then vacuum should be broken by opening the valve connecting turbine exhaust hood to atmosphere and stopping the ejectors.

As the lub oil temperature at bearing exit is indication of performance of bearing vibration, turbine should be stopped for safety of bearing, if this temperature exceeds 75ºC . As the rotor vibrations are excessive near critical speeds, it is imperative that the critical speed be passed as quickly as possible when increasing the rotor speed. For BHEL turbines, the critical speeds of the shaft system are:

1585, 1881, 2017, 2489 and 4900 rpm.

Sharp increase or decrease of speed/load on turbine due to any reasons should not be permitted. In order to avoid any thermal stresses in the turbine, at the time of heating of the casing of ESV,HPT and IPT and also when increasing speed and loading turbine, metal temperature should be raised gradually, avoiding sudden sharp rise. The permissible rate of rise of metal temperature for the turbine is shown in the table 9.1.

Table 9.1: The Permissible Rate of Rise of Metal Temperature for the Turbine

Temperature range	Rate
From 100 to 200º C	20º C per 5 minutes
From 200 to 300º C	15º C per 5 minutes
From 300 to 400º C	10º C per 5 minutes
From 400 to 500º C	10º C per 10 minutes
From 500 to 535º C	6º C Per 10 minutes

The rate of heating of steam pipes should not exceed 25 to 30ºC per five minutes.

Temperature of main steam and hot reheat steam should not have sharp fluctuations. Other points to be taken care of to avoid any thermal stress are:

1. While heating the main steam and reheat steam pipes, the difference in temperature of pipes should not exceed 15ºC But when the turbine is on load, this difference should not be more than 10º C.

2. While raising the speed of set up to 3000rpm the difference of metal temperature between left hand and right hand ESV should not exceed 30ºC.

3. Temperature difference between the upper and lower halves of HPT and IPT casing should not exceed 50ºC near the regulation stage in the case of HPT and near the zone of steam admission in case of IPT.

4. While switching on the heating of flanges and studs, the entry of heating steam should be controlled so that the difference of temperature along the width of flange is not more than 50ºC; the pressure of steam not exceeding 2KG per cm² (abs). The Temperature of outer surface of flanges should be always lower than that of innermost surface of flanges.

5. During the trial runs for the heating of flanges and studs their heating should be adjusted such that:

   a. Temperature difference between upper and lower flanges does not exceed 10ºC.

   b. Temperature difference between left hand and right hand flanges does not exceed 10º C

   c. Temperature difference between flanges and studs does not exceed 20ºC and is not negative, i.e. temperature of stud must be lower than that of flanges.

   d. While starting the turbine under any condition, the metal temperature difference between inner and outer surface of the wall of HPT casing should not exceed 35ºC.

The speed or load of turbine should be raised at such a rate that differential expansion or the relative expansion or contraction of the rotor with casing does not exceed the maximum permissible values of +4.0mm to -1.2mm for HPT, +3.0mm to -2.5mm for HPT and +4.5mm to -2.5mm LPT ('+' sign meaning that rotor is longer than cylinder and vice versa). The differential expansion of rotor beyond permissible limits can be controlled by taking followed by taking following measures:

a. If HP rotor expands quicker than the HPT casing, heating steam supply for heating flanges and studs should be increased ensuring that the pressure ih header supplying steam for heating does not exceed 2KG/sq .cm absolute. In case of further increase in live steam temperature and suspend further loading. If measures mentioned above are not sufficient, decrease the live steam temperature.

b.   If HP rotor contracts quicker than the HP casing stop supply of heating steam to flanges and studs, increase the temperature of live steam and increase load or supply fresh steam to the front sealing of HPT.

c.   Similar actions are re3quired for IPT differential expansions.

d.   In case, LP rotor expands quicker than LPT cylinder, then vacuum should be worsened and recirculation of condensate in the condenser stopped. And Vice versa if LP rotor contracts quicker than LPT casing (vacuum in the condenser should be improved)

9.8. Procedure of Starting Turbine from Cold State

Once the internal super heater flow is established through start up vent, the HP bypass valves are fully opened, thus establishing steam flow through and assist in heating of main steam, cold and hot reheat lines and also increasing the parameters of the steam HPT evacuation valves should be closed. The heating rate of steam lines is controlled within prescribed limits by adjusting the position of Hp bypass valves. After 50ºC super heater is obtained at the superheat outlet, the bypass valves of the main steam stop valve MSV are opened there by permitting heating of the pipe lines up to turbine. After that the emergency stop valves (ESV) are opened and the steam admission pipes up to control valves are heated to a temperature of 150ºC. The HP turbine is also heated up to 150ºC admitting steam into HPT through cold reheat line from the exhaust side of HPT. The steam pressure in hot reheat line should be less than $1Kg/cm^2$ (gauge) before starting the heating of steam admission pipe up to control valves of IPT.

The interceptor valves (IV) are opened and thus the steam admission pipes up to control valves are heated to a temperature of 100ºC. During the process of heating of the steam pipe lines and of turbine, it is very important that the drain valves should be open and there should be no clogged drain. When steam pressure before MSV is 12 to 14 Kg/cm^2 and temperature between 24.ºC and 26ºC and if ESV, steam admission pipe of HPT are heated up to 150ºC, and the interceptor and control valves of MSV ,ESV and IV are closed. Hp bypass is also closed and LP bypass opened to attain condenser pressure in the reheat line after which LP bypass is also closed. If all metal temperature and turbo supervisory measurements are within limits and if condensate level in hot well is within allowable limits, then control valves(CV's) of HPT and IPT are completely opened and the turban is now ready for rolling. The bypass valves of MSV are opened slowly, thereby allowing the steam to roll the turbine. The speed is raised to 500 rpm and turbine soaked for 5-10minutes at this speed and its healthiness in respect of vibrations, noise and rubbing is checked. The speed is then raised to 1200 and a rate of 50rpm/minute by slowly opening the bypass valves of MSV, and the set is soaked at 1200rpm for about 20 minutes.

To ensure uniform heating, studs and flanges of HPT and IPT should be heated, keeping a close watch on permissible metal temperatures. The speed is then smoothly raised to 300rpm without pause, never holding the set in the critical speed zone. The turbine is held at 3000rpm for about 20-25 minutes with a view to carryout inspection, listening and soaking of turbine, testing the protections and governing system, tightness of closure of ESV and control valves of HPT, a close watch on maintaining condenser vacuum, lub oil temperature before IPT around 240-260ºC. Before synchronization, with the help of control gear control valves are closed to such an extent that speed begins to fall, and the MSV is completely opened and by pass valves are closed, and with the help of control gear the speed of the set is maintained at 3000rpm. The set is synchronized then it is very important to nbote that speeding, loading and increasing of parameters should be crried out strictly in conformity with the recommendations of turbine manufacture. Fig 9.2 shows the curve for turbine "starting the turbine from cold state"

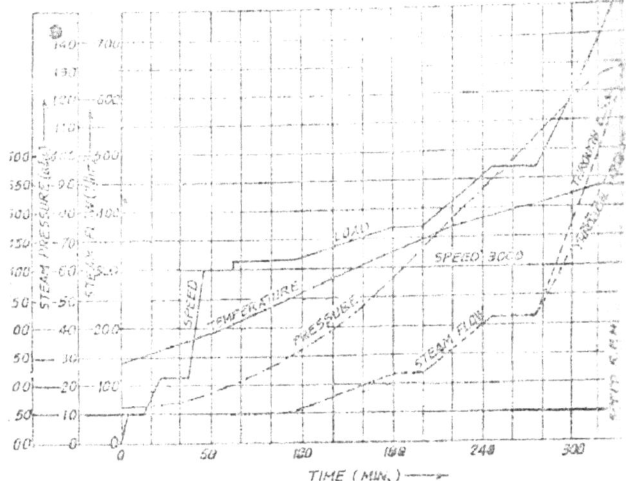

Fig. 9.2: The Curve for Turbine "starting the turbine from cold state"

Measure to avoid entry of water into the steam flow path or in the end sealing of turbine deserve special attention.

9.9. Loading of the Set

After synchronizing the set with the grid, a load or 5-7 MW is taken and the turbine is held at this load for 30 minutes to carry out soaking. After soaking the set of about 5-10 minutes at a load of 5-7 MW, the entire drain valve on steam lines, steam admission pipes high pressure drain collector and intermediate pressure drain collector are closed. It is important to keep a close watch on differential expansion of the rotors during the heating of turbine under load. Starting of the turbine should be so regulated that the temperature difference between outer and inner surface of the wall of HPT casing at regulating stage does not exceed 35ºC. After ensuring the satisfactory and healthy operation of the turbine at the load5-7MW, the load is increased to 30 MW in a period of about 70 minutes, and the set is held at this load for 20minutes for soaking of the turbine. Steam parameters are continuously raised smoothly to attain steam pressure of 61 Kg/cm^2 and temperature of 430ºC at the end of this soaking period. Temperature should be raised smoothly and in no case faster than 4ºC per minute in the range of 400ºC to 535ºC. All the drain cocks in the extraction lines of low pressure heaters are closed now and the load is increased to 70MW in a period of 50 minutes and the set held at this load for 30 minutes for soaking of the turbine. At this time the high pressure heaters are brought in service. After soaking at 70MW, the steam parameters are expected to reach 105Kg/cm^2 and 490ºC. The supply of heating steam go the flanges and studs is stopped after ensuring that the metal temperature has stabilized and the difference of temperature across the width of the flanges is 20-30ºC. The load is smoothly and gradually increased to 210 MW in a period of 70 minutes and simultaneously the steam pressure is raised to the rated values.

9.10. Warm and Hot Start of Turbine

If the metal temperature of the HPT casing in the zone of regulating stage is above 350ºC, at the time of starting turbine it is called hot start, and if the temperature is between 150-350ºC, it is known as warm start. In the case of warm/hot start also, the auxiliary equipment should be started first as in case of cold start up and the necessary precaution in respect of temperature limits taken. Steps should be taken to ensure that lub oil, governing oil pressure, steam supply to end sealings, condenser vacuum and condenser levels are established. Steam lines should be heated after the vacuum in condenser is 600 mm of HG, keeping MSV, ESV, CVs, or HPT and IPT tightly closed. Heating of the fresh steam lines, body of ESV, reheat pipe lines, body of IV may be done in case the temperature of these elements is 80-100ºC less than the temperature of the lower half of the HPT casing in the Zone of regulating stage and in the zone of steam admission in case of IPT respectively. All the drains of relevant should be opened before starting the heating of steam pipelines. More effective heating of steam pipelines and raising of steam parameters to match the turbine requirements can be achieved by putting HP by pass into service. In case of shut down for less than 6 hours, steam admission pipes of HPT and IPT bodies of ESV and IV do not get cooled significantly and, therefore prior heating of these elements is not required and it is sufficient to blow these lines for 5 minutes through drains before rolling.

Heating of ESV body and steam admission pipe from ESV up to CV of HPT, if necessary, is done with steam which is 50ºC hotter than the metal temperature of the hottest part of turbine ESV, by opening gradually the bypass valves of MSV and keeping ESV open and control valves completely closed . Heating is done upto the temperature existing in the zone of regulating stage of HPT. Heating of reheat lines is accomplished by allowing fresh steam through HP by pass station. The temperature of steam entering the cold reheat line should be about 50ºC higher than the metal temperature of cold reheat line, but not more than 380ºC. Before admitting steam to the turbine is rotating on barring gear, shaft eccentricity should be checked and the difference between upper and lower halves of HPT and IPT should not be more than 50ºC

9.11. Warm Start

The desired steam temperature and pressure for rolling the turbine can be determined on the basis of metal temperature matching charts supplied by manufacturer. However the steam temperature must be more(by least 25ºC) than the hottest metal temperature of ESV and steam admission pipes of HPT. HP bypass valves is opened and steam parameters of boiler are raised to desired matching value, and simultaneously heating the main steam, cold reheat and hot reheat line. The steam admission pipes are also heated up to control valves by opening by pass valves of main stop valves. All drain lines are kept open and checked against clogging. The steam pressure in hot reheat line should be kept less than $1Kg/cm^2$. After desire steam temperature is attained the HP by pass valves at ESV, are closed and the LP bypass valves are opened. When the pressure in the reheat line comes to condenser vacuum the LP bypass valves are closed. With the help of the speeder gear, ESV, IV and CVs of HP and IP turbines are fully opened, and the bypass valves of MSV are slowly opened and the turbine is rolled and the speed is raised to 500rpm. Vibrations of the various bearings are checked. Being certain that the turbine is in healthy state, the speed is raised to 1200rpm within 3-4minutes by further opening bypass valves of MSV, and the set is soaked at this speed for 2-3 minutes. The speed is then raised to 3600 rpm without pause, and held at 3000 rpm for 5minutes with a view to carry out the inspection checking rubbing and soaking of turbine. The set is synchronized and load or upto 20-30MW is applied. All the drains are closed after some load has been taken on the turbine. The steam parameters and loading of the set is carried on as per the curve for the warm start

9.12. Hot Start

As for warm start, the parameters of the steam are raised to get the desired matching temperature with the help of HP-LP bypass station. The temperature at HP bypass exit is controlled to meet then boiler requirement and ensuring the allowed heating rate of cold reheat pipe lines are repeater. The entrapped steam from HP casing is released by opening the evacuation valves connecting HP casing to condenser. Before rolling the rotor, fresh steam is supplied to from sealing of HPT and IPT if the contraction of HPT rotor has reached 0.8mm and that of IP rotor 1.5 mm. Before supplying the fresh steam to front seals of HPT and IPT the concerned pipe lines should be adequately heated up, and blown out. The transfer pipe between IV and CV of IPT are heated by opening IV keeping the valves in the drains of these pipes fully open. After heating these pipes corresponding to the available steam temperature, IV is fully closed. MSV are fully opened and also the valves in the drain lines of the transfer pipes are fully opened. ESV and Iv are also fully opened with the help of the control Gear. Turbine will star rolling through IPT. CVs and IPT are opened and the turbine speed is raised to 3000 rpm without pause. The set is synchronized and load around 20 MW load is taken. All the drains are closed. The pressure in reheat line should be controlled at around 6ata. The turbine is loaded at the rate of 5MW per minute until the Hp bypass gets closed under the pressure and the set is held at this load. The steam starts flowing through HP turbine. Further loading of the set is done as per the curve for hot start by increasing boiler firing rate. The time for loading these during hot restart depends on the thermal condition of HPT and IPT. While loading the turbine it should be ensured that the metal temperature difference between inner and outer surface of the wall of HPT casing does not exceed 35ºC. Otherwise further loading of the set should be stopped until this requirement is met. Fresh steam supply to the sealing's of HPT and IPT is stopped when different expansion of HPT and IPT rotors has reduced and has got established.

9.13. Planned Shut Down Of Turbine

Before proceeding to shutdown turbine, one should check the starting oil pump and lubricating oil pumps and ensure automatic starting of lubricating oil pumps, check the non-seizure of ESV and IV, and that bypass valves of MSV are closed. The turbine is unloaded by closing gradually the control valves through remote or manual operation of control gear, at a rate of 3MW per minute at rated parameters. The rate of unloading mainly depends on differential contraction of HP rotor. When load is reduced to 160MW, the HP heaters are switched off. One of the fed pumps and one condensate pump are switched off at load of 90-100MW. Sharp fall in metal temperature of steam distribution elements of HPT and IPT should be avoided and all turbovisory instruments watched very closely. If there is jamming in ESV or IV, MSV should be closed immediately. When the speed has dropped to 2800 rpm the AC lubrication oil pump should be started. When rotors come to a stop immediately the barring gear should be started and the rotors rolled continuously till temperature of lower part HP casing at regulating stage drops to 170ºC. After rolling the rotors on barring gear, vacuum is broken and the steam supply to the sealing is stopped. The condensate pumps are stopped and the condensate supply to pump sealings and the cooling water to interceptor and emergency stop valve servomotors are also stopped. After the turbine is shut down, drain and blow down cocks on steam admission pipes and cylinders are not opened till the metal temperature falls below 200ºC . If the turbine is to be started before the metal temperature drips to 200ºC, then drain and blow down cocks are only opened during preparation for restarting. Drains should remain closed if the turbine has been stopped for a period not exceeding 6 hours. Circulating water pumps should be stopped only when the temperature of exhaust part of LPT falls to 55ºC. If the turbine has to be shut down for 6-8hours suitable measures   should be taken so that the steam admission parts get cooled to the minimum.

9.14. Common Turbine Troubles

The balding, clearances and bearings and alignment are usually the most sensitive areas of turbine. Turbine performance is very much affected by the condition of balding which is subjected to deposition and erosion problems. Sometimes foreign matters with steam or gulps of water from boiler priming cause lot of mechanical damage. Deposits on blades are usually formed due to NaOH (Soluble), or silica(Insoluble) carried over from the boiler which may be either of mechanical nature, i.e. minute particles swept with may be either of mechanical nature, i.e. minute particles swept with steam or may result from volatilization of silica at pressures above 40Kg/cm². Thus the boiler water to be kept to minimum. At lower temperatures, i.e. in later stages of turbine NaOH forms and sticky mass on blades which helps the particles of sulphates and phosphates and silica to deposit further on blades. The deposits on blades lower down the efficiency and capacity of turbine. Excessive deposits can lead to reverse thrust loads thereby damaging the thrust bearings/blades stripping.

Some deposits are removed during cold start up of turbine due to washing action of moist steam and cracking off of silica due to the sudden temperature changes. The insoluble deposits are removed by chipping off with suitably shaped chisels. Deposits can also be removed by washing process. It consists of lowering the steam temperature by about 55ºc per hour with the machine on load. When temperature has dropped to a tolerance limit (about 75% of normal) the machine is off loaded and slowed to one-fourth normal speed by closing stop valve suitably and keeping throttle valves fully open. At this stage water is injected at a rate such the cooling rate approximates to the rate set out above and continued till pure condensate is received. The machine is then speeded up and load raised. During washing a careful watch on differential expansion and rubbing should be kept. Sometimes. Silica is removed by immersing rotor in hydrofluoric acid with suitable inhibitors. Sometimes caustic soak and steam washing is also employed for removal of silica.

9.15. Performance and Testing of Turbine

The following types of tests may be carried out on a turbine:

1. **Full scale testing. Acceptance tests**: This is done by turbine manufacturer soon after commissioning to demonstrate the requirements of specifications. Many special instruments of best accuracy(laboratory type other than the conventional instruments for routine operation are employed).
2. **Routine Testing**: Its purpose is to check heat consumption of turbine.
3. **Part tests for special purpose**: These are used to check performance of equipment like heaters, condenser, ejector, stage pressures for gauging blade deposits etc.

In routine tests, the calculation are made for heat rate(overall), turbine output, steam consumption, thermal efficiency, Ranking cycle, efficiency ratio, stage efficiency etc.

9.16. Testing of Turbine

The procedure described below is not intended for turbine acceptance tests or determining the absolute level of performance but for the analysis and supervision of relative performance of turbine throughout its service life time, i.e. to monitor changes tests will be of use in providing guidance in scheduling maintenance outages on the basis of performance; providing guidance in establishing the loading sequence of steam turbine-generator units according to their relative performance; evaluating major modification of the turbine or turbine cycle, and changes in operating procedures, detecting performance changes in specific areas of the turbine or turbine cycle.(This turbine test is based on a report by ASME performance test code committee no:6)

Definitions of the terms used in test are:

1. Steam Rate

Steam consumption per hour per unit output (Kg per KW hr) in which the turbine is charged with the net steam quantity supplied, with correction for pickings, if necessary.

2. Heat Rate

Heat consumption per hour per unit output (Kcal per KW hr). In this the turbine is charge with the aggregate enthalpy (Product of enthalpy, kcal per kg, and rate of flow, kg per hour) of the steam supplied to the throttle plus the aggregate enthalpy added by the repeaters. It is credited with the aggregate enthalpy of the feed water leaving the final heater. Corrections may be required for

- Temperature of the condensate if below that corresponding to the exhaust pressure
- Internal losses and theoretical work of any pumps between the condenser and steam generator
- Other aggregate enthalpy entering leaving cycle, if necessary

3. Valve Point

The valve position just before the succeeding valve starts to open.

4. Mean of the Valve Loop

A smooth curve which gives the same load-weighted average performance as the valve-loop curve.

5. Enthalpy Drop

The difference in enthalpy between steam at the turbine inlet condition and at turbine outlet conditions. This is applicable to individual turbine sections such as HP section or IP section.

9.16.1. Advance Planning

Advance planning for the test is essential. A plan for instrumentation of the turbine and turbine heat cycle should be developed prior to installation.

The various items to be considered are:

1. Location and installation of a calibrated primary flow metering section.
2. Provision for the accurate measurement of output.

3. Location and installation of test connections for primary pressure and temperature measurements.

4. Provision for measurements of secondary leak-off any by-pass flows which may affect the primary–flow measurement or have a significant effect in calculation of test performance.

5. Select of test instruments capable of the repeatability required for consistent test results.

6. Location of test instruments in groups to facilitate calibration and use, and minimize the number the observation required.

9.16.2. Turbine Cycle Isolation

Before conducting test, it is essential to make a water balance of the cycle by positive isolation of the turbine cycle. This establishes the magnitude of unaccounted-for leakage which directly affects the accuracy of the test result. A satisfactory water balance of the cycle should show an unaccounted for loss of less than ± 0.5% of the maximum turbine throttle flow to provide acceptable test accuracy. Any change in the unaccounted for loss exceeding 0.1% of the maximum throttle flow should be investigated, since it may directly affect the test result. If the cause can be determined, then the test result should be adjusted to compensate for this factor. Some pertinent locations in this regard are

1. Hot well make up and draw off
2. Auxiliary steam lines for the boiler
3. Boiler blow down lines
4. Cross-connections to other steam or water systems.

Minor sources like heater-cycle drains, pump-shaft leakages, feed water sampling flows, and atmospheric vents must also be measured or sampled to make a complete water balance of the cycle.

9.16.3. Preliminary Tests

These include:

1. To establish the location of turbine valve point
2. To isolate the turbine cycle and make a water balance of the cycle.
3. To check internal cycle conditions. Variations in equipment operation such as following, that affect turbine stage flows or stage pressures should not be permitted or should be the subject of an appropriate correction.
   a. Change in feed water heater performance, or bypassing of heater, or

b. Leakage in heaters.

c. Make up evaporator in service.

d. Amount of desuperheating water in use for control of steam temperature

4. To check the operation of test instrumentation and familiarizes observers with the routine of proper testing of procedures.

5. Calibration of test instruments should establish prior to test run

The duration of the test run should be adequate to establish accurate average data and accurate integrated primary measurements. If possible, turbine should be tested with governing valves wide open. Test conditions should be established and maintained which are as close to specified conditions as possible to minimize the magnitude of test correction-factor adjustments to the turbine test heat rate. The recommended reading frequencies for test readings are as follows.

1. Each 2 minutes-primary flow, power output turbine speed.

2. Each 10 minutes-secondary flows affecting throttle flow calculation, primary pressures and temperatures.

3. Each 15 minutes-secondary flows not affecting throttle flow calculation.

4. Each 20 minutes- secondary electrical data not affecting generator output calculation, secondary pressure and temperatures.

5. Each 30 minutes- integrated flow integrated electric power measurements, storage level changes.

Recording test data should be examined for consistency and reliability. Inconsistent readings may be discarded provided at least 90% of the total readings are retained for averaging. The following steps may be followed for data reduction:

1. Average all test readings and obtain meter integrated flow differences and generator-output differences.

2. Apply water leg and gauge calibrations to averaged pressure readings and convert pressure readings into Kg/cm^2 abs. value.

3. Convert thermocouple output to temperature units and apply thermocouple calibration.

4. Apply meter calibrations and working fluid density corrections to integrated quantities.

5. Convert storage-level difference into Kg/hr.

6. Correct generator output to specified power factor and hydrogen pressure.

7. Calculate primary and secondary flows.

8. Calculate steam and water enthalpies.

9.16.4. Performance Test on Turbines

It follows the recommendations of BS 752-1974 in conducting the performance test on their turbine. All important measurements essential for verifying the guaranteed specific heat are made.

Before conducting the test, the water tightness of condenser, regenerative heaters is ensured. The air tightness of the condenser is determined by vacuum drop of not more than 2mm mercury/min. 80% load with the ejectors being shut down. During the guarantee measurement the supply of feed make up water to condenser and recirculation devices are kept closed. All the auxiliary connections of the water or steam which may influence the test result but are not part of the basic scheme are closed. The maximum permissible deviation of the average test condition for each variable from the specified and the maximum possible fluctuation of the variable during any one test run should not exceed the limit prescribed below.

Table 9.2: The Maximum Permissible Deviation of the Average Test Condition for Each Variable from the Specified and the Maximum Possible Fluctuation of the Variable During any One Test Run

Variable	Maximum permissible deviation of the average of the test from specified value	Maximum permissible rates of fluctuation from the average during any one test run
Initial steam pressure	± 3% of absolute pressure	± 0.5% of absolute pressure
Initial steam temperature	Superheat<52± 8º C	±2ºC
Reheat steam temperature	Superheat <25± 15º	±4ºC
Exhaust pressure	+25% of absolute pressure -10% of absolute pressure	±5%
Final Feed water temperature	±8ºC	--
Power	±5%	±0.25%
Voltage	±5%	--
Power factor	Play vary between unity and	___
Cooling water inlet temperature	(specific value -.0.05) ±5ºC	±1ºC
Cooling water flow	±10%	_

Correction on the heat rate due to variation in the operating conditions is done in accordance with the test correction curves. All the instruments are calibrated immediately before the test. All instruments except flow element are recalibrated immediately after the test. As the measurement of each quantity entering into the computation of the test results is liable to some degree of error the test result is subject to uncertainty depending on the combined effect of all the errors of measurement. According an extra tolerance is put in the calculated heat rate to compensate the error in calculated heat rate due to errors of measuring instruments. The amount of this tolerance is estimated by taking the square root of the sum of square of all the separately calculated errors in the heat rate due to each error of measurement.

The temperature at various points in the cycle above 500ºC is measured by calibrated high grade thermocouple and precision potentiometer: the cold end of the thermocouple being immersed in the ice water mixture with minimum quantity of ice being 50%. For temperature up to 500ºC, Calibrated electrical resistance thermometer with a precision bridge is used. Temperature measurement is made at points as close as possible to the points at which the corresponding pressures are measured. Pressure is measured by means of Bourdon tube type pressure gauge, or with U tube manometers. The pressure gauges are calibrated before and after the test. The pressure drop in stop valves, governor valve, and reheat circuit are measured by means of differential bourdon tube type gauges. Absolute condenser pressure is measured with the help of Baro-vaccum meters.

The various steam and water flows are measured by measuring differential pressure across the test/plant measuring orifice plates/checked before the erection. Pressure and temperature of medium should also be measured near the flow measuring device to find the specific weight of the flowing medium.

9.17. Condensers and Cooling Systems

9.17.1. Condensers

The function of the condenser is to condense exhaust steam from the steam turbine by rejecting the heat of vaporization to the cooling water passing through the condenser. The temperature of the condensate determines the pressure in the steam/condensate side of the condenser. This pressure is called the turbine backpressure and is usually a vacuum. Decreasing the condensate temperature will result in a lowering of the turbine backpressure. Note: Within limits, decreasing the turbine backpressure will increase the thermal efficiency of the turbine.

The condenser also has the following secondary functions:

- The condensate is collected in the condenser hot well, from which the condensate pumps take their suction
- Provide short-term storage of condensate

- Provide a low-pressure collection point for condensate drains from other systems in the plant; and

- Provide for de-aeration of the collected condensate.

A typical power plant condenser has the following functional arrangement.

Large power plant condensers are usually 'shell and tube' heat exchangers. These types of condensers are also classified:

- As single pressure or multi-pressure, depending on whether the cooling water flow path creates one or more turbine backpressures;

- By the number of shells (which is dependent on the number of low-pressure turbine casings); and

- As either single pass or two-pass, depending on the number of parallel water flow paths through each shell.

Other Types of Condensers Are

- Plate types consisting of a series of parallel plates that provide paths for the steam and the cooling water. Plate condensers are used mainly for smaller power plants; and

- Direct contact types where the cooling water is sprayed directly into the steam. This type of condenser is used in applications where the cooling water is the same quality as the steam condensate. Systems that have dry cooling (described in a following section) sometime use direct contact condensers.

The parts of shell and tube condensers and plate condensers involved in the transfer of heat from the steam and condensate to the cooling water should have the following properties:

- Be resistant to corrosion from both the steam/condensate and the cooling water;

- Have a minimal resistance to the flow of heat from the steam/condensate through the material into the cooling water; and

- Provide mechanisms to remove organic and inorganic deposits on the heat transfer surfaces in contact with the cooling water.

9.18. Types of Cooling Systems

Some power stations have an open cycle (once through) cooling water system where water is taken from a body of water, such as a river, lake or ocean, pumped through the plant condenser and discharged back to the source. Inland plants away from large water bodies prefer to use closed cycle wet cooling system with wet cooling towers. Plants in remote dry areas without economic water supplies use closed cycle dry cooling systems that do not require water for cooling. Hybrid cooling systems are used in particular circumstances.

The type of cooling system used is therefore heavily influenced by the location of the plant and on the availability of water suitable for cooling purposes. The selection process is also influenced by the cooling systems environmental impacts (refer to a following section for a brief discussion on this topic).

9.18.1. Open Cycle Cooling Systems

Open cycle (once through) cooling systems may be used for plants sited beside large water bodies such as the sea, lakes or large rivers that have the ability to dissipate the waste heat from the steam cycle. In the open system, water pumped from intakes on one side of the power plant passes through the condensers and is discharged at a point remote from the intake (to prevent recycling of the warm water discharge).

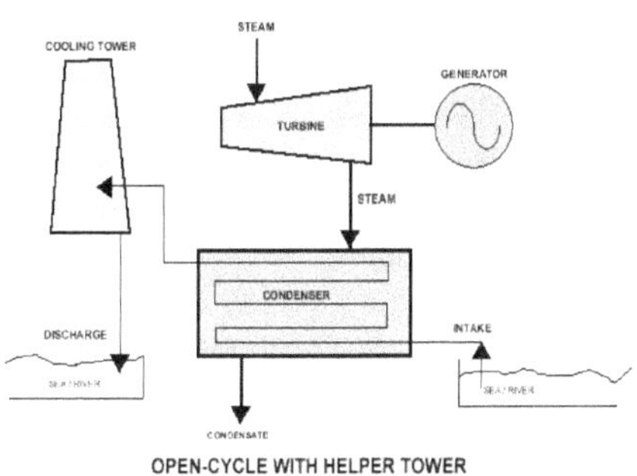

Fig. 9.3: Open Cycle with Helper Cooling Tower

Open systems typically (shown in figure 9.3) have high flow rates and relatively low temperature rises to limit the rise in temperature in the receiving waters. A typical 350 megawatt (MW) unit would have a flow of some 15,000 litres per second (L/s) to 20,000 L/s.

Lake cooling systems are a variant on a true open system as the temperature of the lake is increased from the circulation of the warm water. Environmental requirements have become more stringent on the allowable rise in temperature of the receiving waters, so that closed systems are now more commonly used in Australia.

9.18.2. Open Cycle with Helper Cooling Tower

In this system, cooling towers are installed on the discharge from open systems in order to remove part of the waste heat, so that the load on the receiving waters is contained within pre set limits. Systems with helper cooling towers are common in Germany and France where cooling supplies are drawn from the large rivers. The helper towers are used in the warmer summer periods to limit the temperature of the discharged cooling water, usually to less than 30ºC.

9.18.3. Closed Cycle Wet Cooling Systems

In closed cycle wet cooling systems, the waste energy that is rejected by the turbine is transferred to the cooling water system via the condenser. The waste heat in the cooling water is then discharged to the atmosphere by the cooling tower.

In the cooling tower, heat is removed from the falling water and transferred to the rising air by the evaporative cooling process. The falling water is broken up into droplets or films by the extended surfaces of the tower 'fill'. This 'fill' in the later Queensland towers is manufactured from plastic. Some of the warm water, typically 1% to 1.5% of the cooling water flow, is transferred to the rising air, and this is visible in the plume of water vapour above towers in times of high humidity as shown in the figure 9.4. The evaporation rates of the Queensland 350 MW cooling systems are typically 1.8 litres of water per kilowatt hour of power generated.

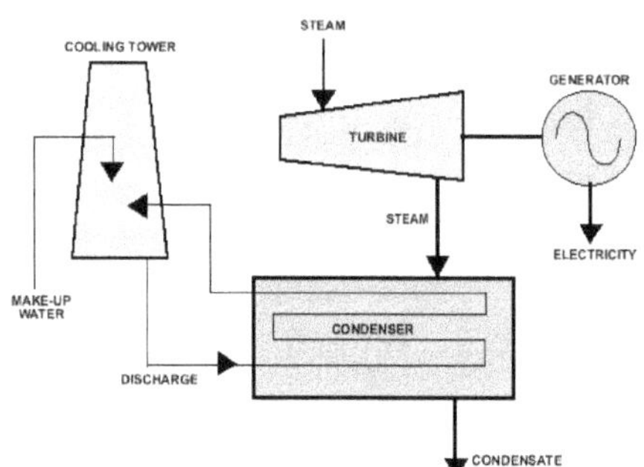

Fig. 9.4: Closed Cycle Wet Cooling Systems

The major components of a closed cycle wet cooling water system are:

- Cooling towers - two types are commonly used, concrete natural draught towers and mechanical draught towers; and
- Pumps and pipes.

9.19. Natural Draught Towers

Concrete natural draught towers have a large concrete shell. The heat exchange 'fill' is in a layer above the cold air inlet at the base of the shell as shown in the tower sectional view. The warm air rises up through the shell by the 'chimney effect', creating the natural draught to provide airflow and operate the tower. These towers therefore do not require fans and have low operating costs.

The cooling towers have two basic configurations for the directions of the flow of air in relation to the falling water through the tower fill:

- The counter-flow tower where the air travels vertically up through the fill (a diagram of this type of tower is shown below); and
- The cross-flow tower where the air travels horizontally through the fill.

Natural draught towers are only economic in large sizes, which justify the cost of the large concrete shell. Natural draught towers are the most common towers for large generating units in Europe, South Africa and eastern United States of America. They are not used in the drier areas of western United States of America, as their performance is better suited to cooler and more humid areas as shown in figure 9.5.

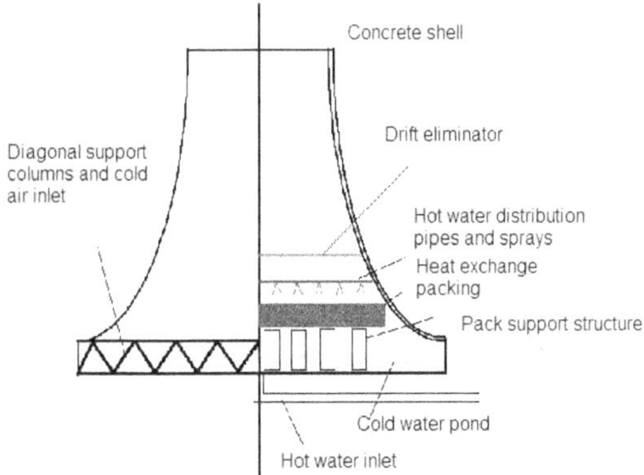

Fig. 9.5: Part Section of a Draught Cooling Tower-Counter flow

9.20. Mechanical Draught Cooling Towers

In mechanical draught cooling towers, large axial flow fans provide the airflow. While fans have the disadvantage of requiring auxiliary power, typically 1.5 MW to 2.0 MW for a 420 MW unit, fans have the advantage of being able to provide lower water temperatures than natural draught towers, particularly on hot dry days. Mechanical draught towers are used exclusively in central and western United States of America as their climate can vary from freezing to hot with low humidity, and the mechanical towers can provide a more controlled performance over this wide range of conditions. The most common materials used in large mechanical draught cooling towers are timber for the framing and plastic for the cladding and internals.

9.21. Pumps and Pipes in a Cooling Water System

Circulating water pumps supply cooling water at the required flow rate and pressure to the power plant condenser and the plant auxiliary cooling water heat exchangers. These pumps are required to operate economically and reliably over the life of the plant. The three types of pumps commonly used for circulating water service are 'vertical wet pit', 'horizontal dry pit' and 'vertical dry pit'. For once through systems, vertical wet pit pumps are in common usage. For re-circulating cooling systems, vertical wet pit and horizontal dry pit are used about equally, with occasional use of vertical dry pit pumps. Circulating water piping carries the cooling water from the circulating water pumps to the condenser and returns the water to the cooling tower or discharge structure. The large flow rates associated with circulating water systems typically require the use of large diameter piping in the range 900 mm to 2400 mm diameter. The design of the pipe work must consider the environment internal to the pipe as well as the external environment. Pipe materials used include steel, fibre reinforced plastic and reinforced concrete. The large water requirement generally makes it uneconomical to use high quality water sources. The source of water for the plant generally depends on the plant's location. Coastal sites generally use seawater or brackish water as the circulating water source, either by pumping directly from the sea or extracting the water from the local bores. Water from many sources can contain high concentrations of corrosive contaminants. Any pipe materials considered must include measures to protect the pipe for the service life of the plant. For example, carbon steel pipes in seawater service require either an internal coating, or a cathodic protection system, or both. Concrete pipes may require a dense concrete mix to withstand chloride attack. These protective measures significantly increase the capital cost of an installation such that it can be as economical to install fibre reinforced plastic pipe to obtain the same service life. As existing water sources become strained and new water sources more scarce and expensive to develop, the quality of circulating water in future power plants is expected to decline further. This will increase the trend towards corrosion resistant piping materials.

9.21.1. Closed Cycle Dry Cooling Systems

Dry cooling systems are used where there is insufficient water, or where the water is too expensive to be used in an evaporative system. Dry cooling systems are the least used systems as they have a much higher capital cost, higher operating temperatures, and lower efficiency than wet cooling systems.

In the dry cooling system, heat transfer is by air to finned tubes. The minimum temperature that can be theoretically provided is that of the dry air, which can be regularly over 30ºC and up to 40ºC on typical summer afternoons in Queensland.

Compare this to wet cooling towers, which cool towards the wet bulb temperature, which is typically 20ºC on summer afternoons. The steam condensing pressures and temperatures of a dry cooled unit are significantly higher than a wet cooled unit, due to the low transfer rates of dry cooling and operation at the dry bulb temperature.

There are two basic types of dry cooling systems:

- The direct dry cooling system; and
- The indirect dry cooling system.
- Variations on the full dry and full wet systems are hybrid systems, which may be wet with some dry or dry with part wet.

9.21.2. Direct Dry Cooling System

In the direct dry system (shown in figure 9.6), the turbine exhaust steam is piped directly to the air-cooled, finned tube, condenser. The finned tubes are usually arranged in the form of an 'A' frame or delta over a forced draught fan to reduce the land area. The steam trunk main has a large diameter and is as short as possible to reduce pressure losses, so that the cooling banks are usually as close as possible to the turbine.

The direct system is the most commonly used as it has the lowest capital cost, but significantly higher operating costs. The power required to operate the fans of this system is several times that required for wet towers, being typically 4 MW to 5 MW for a 420 MW unit.

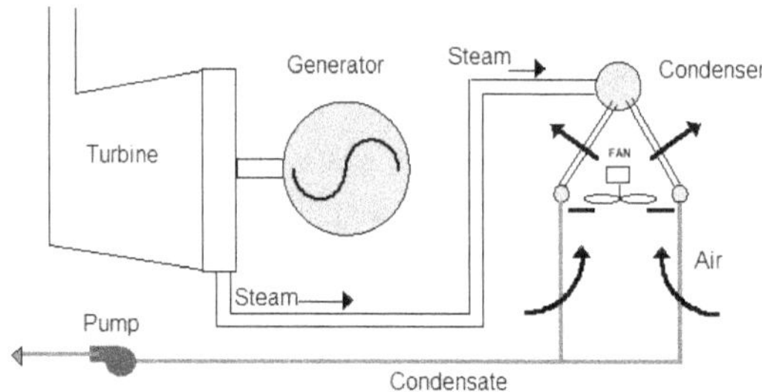

Fig. 9.6: Direct air Cooling System

9.21.3. Indirect Dry Cooling System

Indirect dry cooling systems have a condenser and turbine exhaust system as for wet systems, with the circulating water being passed through finned tubes in a natural draught cooling tower. The water pipe work allows the towers to be sited away from the station as shown in figure 9.7.

A variation on this type of indirect system is the system that uses a direct contact condenser in place of the traditional tube type condenser. In the spray condenser, the water from the cooling cycle mixes with the boiler water. The maintenance of the water quality to suit all circuits is critical to the successful operation of the system.

Hybrid Systems

There are two common hybrid systems, which have been developed to overcome some of the disadvantages of the full wet and full dry systems.

Wet with Part Dry

One of the problems with wet towers is that in cold and humid climates the towers plume can create fog. In the part dry or plume abatement tower, a dry section above the wet zone provides some dry cooling to the exhaust plume to remove the condensing water vapour. These towers are common in Germany and England where environmental problems with mechanical towers have arisen.

Dry with Part Wet

Problems with full dry towers are centered on loss of performance in hot weather. With the part wet towers, there is provision for water sprays to evaporative cool the finned tubes for short periods of extreme temperature.

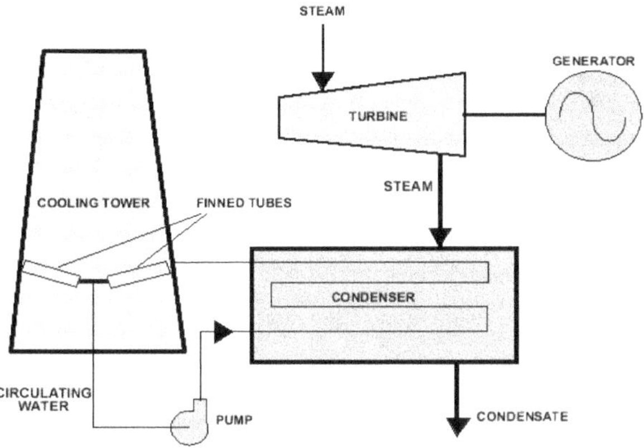

Fig. 9.7: Indirect Dry Cooling System

9.22. Components of Cooling Tower

The basic components of an evaporative tower are: Frame and casing, fill, cold water basin, drift eliminators, air inlet, louvers, nozzles and fans.

9.22.1. Frame and Casing

Most towers have structural frames that support the exterior enclosures (casings), motors, fans, and other components. With some smaller designs, such as some glass fiber units, the casing may essentially be the frame.

9.22.2. Fill

Most towers employ fills (made of plastic or wood) to facilitate heat transfer by maximising water and air contact. Fill can either be splash or film type. With splash fill, waterfalls over successive layers of horizontal splash bars, continuously breaking into smaller droplets, while also wetting the fill surface. Plastic splash fill promotes better heat transfer than the wood splash fill. Film fill consists of thin, closely spaced plastic surfaces over which the water spreads, forming a thin film in contact with the air. These surfaces may be flat, corrugated, honeycombed, or other patterns. The film type of fill is the more efficient and provides same heat transfer in a smaller volume than the splash fill.

9.22.3. Cold Water Basin

The cold water basin, located at or near the bottom of the tower, receives the cooled water that flows down through the tower and fill. The basin usually has a sump or low point for the cold water discharge connection. In many tower designs, the cold water basin is beneath the entire fill.

Some forced draft counter flow design; however, the water at the bottom of the fill is in channeled to a perimeter trough that functions as the cold water basin. Propeller fans are mounted beneath the fill to blow the air up through the tower. With this design, the tower is mounted on legs, providing easy access to the fans and their motors. The types of cooling towers are shown in figure 9.8.

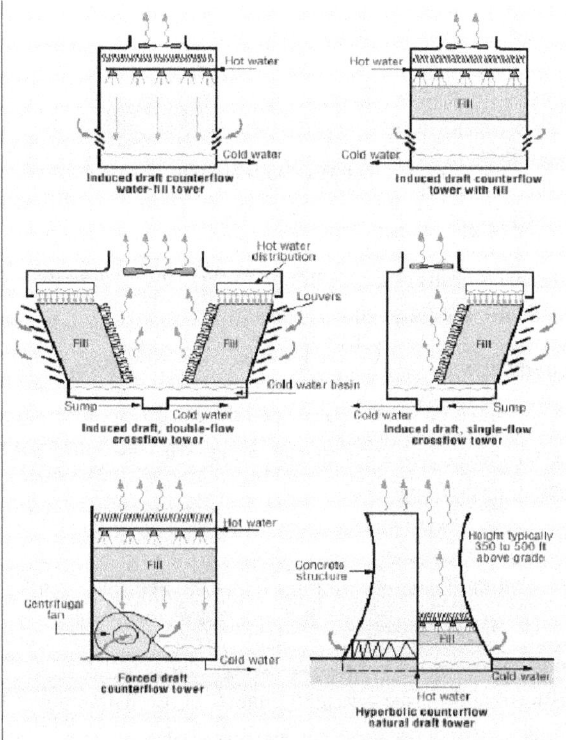

Fig. 9.8: Types of Cooling Tower

9.22.4. Drift Eliminators

These capture water droplets entrapped in the air stream that otherwise would be lost to the atmosphere.

9.22.5. Air Inlet

This is the point of entry for the air entering a tower. The inlet may take up an entire side of a tower-cross flow design-or be located low on the side or the bottom of counter flow designs.

9.22.6. Louvers

Generally, cross-flow towers have inlet louvers. The purpose of louvers is to equalize air flow into the fill and retain the water within the tower. Many counter flow tower designs do not require louvers.

9.22.7. Nozzles

These provide the water sprays to wet the fill. Uniform water distribution at the top of the fill is essential to achieve proper wetting of the entire fill surface. Nozzles can either be fixed in place and have either round or square spray patterns or can be part of a rotating assembly as found in some circular cross-section towers.

9.22.8. Fans

Both axial (propeller type) and centrifugal fans are used in towers. Generally, propeller fans are used in induced draft towers and both propeller and centrifugal fans are found in forced draft towers. Depending upon their size, propeller fans can either be fixed or variable pitch. A fan having non-automatic adjustable pitch blades permits the same fan to be used over a wide range of kW with the fan adjusted to deliver the desired air flow at the lowest power consumption. Automatic variable pitch blades can vary air flow in response to changing load conditions.

9.23. Tower Materials

In the early days of cooling tower manufacture, towers were constructed primarily of wood. Wooden components included the frame, casing, louvers, fill, and often the cold water basin. If the basin was not of wood, it likely was of concrete. Today, tower manufacturers fabricate towers and tower components from a variety of materials. Often several materials are used to enhance corrosion resistance, reduce maintenance, and promote reliability and long service life. Galvanized steel, various grades of stainless steel, glass fiber, and concrete are widely used in tower construction as well as aluminum and various types of plastics for some components.

Wood towers are still available, but they have glass fiber rather than wood panels (casing) over the wood framework. The inlet air louvers may be glass fiber, the fill may be plastic, and the cold water basin may be steel. Larger towers sometimes are made of concrete. Many towers-casings and basins-are constructed of galvanized steel or, where a corrosive atmosphere is a problem, stainless steel. Sometimes a galvanized tower has a stainless steel basin. Glass fiber is also widely used for cooling tower casings and basins, giving long life and protection from the harmful effects of many chemicals. Plastics are widely used for fill, including PVC, polypropylene, and other polymers. Treated wood splash fill is still specified for wood towers, but plastic splash fill is also widely used when water conditions mandate the use of splash fill. Film fill, because it offers greater heat transfer efficiency, is the fill of choice for applications where the circulating water is generally free of debris that could plug the fill passageways. Plastics also find wide use as nozzle materials. Many nozzles are being made of PVC, ABS, polypropylene, and glass-filled nylon. Aluminium, glass fibre, and hot-dipped galvanized steel are commonly used fan materials. Centrifugal fans are often fabricated from galvanized steel. Propeller fans are fabricated from galvanized, aluminium, or moulded glass fibre reinforced plastic.

9.24. Cooling Tower Performance

The important parameters, from the point of determining the performance of cooling towers (shown in figure 9.9) are:

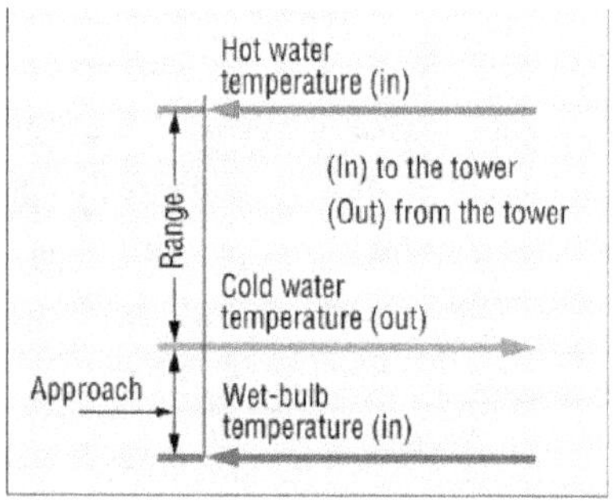

Fig. 9.9: Cooling Tower Performance

- "Range" is the difference between the cooling tower water inlet and outlet temperature
- "Approach" is the difference between the cooling tower outlet cold water temperature and ambient wet bulb temperature. Although, both range and approach should be monitored, the 'Approach' is a better indicator of cooling tower performance.

- Cooling tower effectiveness (in percentage) is the ratio of range, to the ideal range, i.e., difference between cooling water inlet temperature and ambient wet bulb temperature, or in other words it is = Range/(Range + Approach).
- Cooling capacity is the heat rejected in kCal/hr or TR, given as product of mass flow rate of water, specific heat and temperature difference.
- Evaporation loss is the water quantity evaporated for cooling duty and, theoretically, for every 10, 00,000 kCal heat rejected, evaporation quantity works out to 1.8 m3. An empirical relation used often is:

  Evaporation Loss (m3/hr) = 0.00085 x 1.8 x circulation rate (m3/hr) x (T1-T2)

  T1-T2 = Temp. Difference between inlet and outlet water.
- Cycles of Concentration (C.O.C) is the ratio of dissolved solids in circulating water to the dissolved solids in makeup water.
- Blow down losses depend upon cycles of concentration and the evaporation losses and is given by relation:
- Blow Down = Evaporation Loss / (C.O.C. – 1)
- Liquid/Gas (L/G) ratio, of a cooling tower is the ratio between the water and the air mass flow rates.

Against design values, seasonal variations require adjustment and tuning of water and air flow rates to get the best cooling tower effectiveness through measures like water box loading changes, blade angle adjustments. Thermodynamics also dictate that the heat removed from the water must be equal to the heat absorbed by the surrounding air:

$$L(T_1 - T_2) = G(h_2 - h_1)$$

$$\frac{L}{G} = \frac{h_2 - h_1}{T_1 - T_2}$$

Where,

L/G = liquid to gas mass flow ratio (kg/kg)

T1 = hot water temperature (0C)

T2 = cold water temperature (0C)

h2 = enthalpy of air-water vapour mixture at exhaust wet-bulb temperature

(Same units as above)

h1 = enthalpy of air-water vapour mixture at inlet wet-bulb temperature (same units as above)

9.25. Environmental Effects of Cooling Systems

All the heat transferred from the exhaust steam to the cooling system eventually finds its way into the earth's atmosphere. In the once-through cooling water system, heat is removed from the steam turbine and transferred to the source body of water. The heat is then gradually transferred to the atmosphere by evaporation, convection and radiation. However, this waste heat transfer process may negatively affect the body of water buy increasing the temperature of the water.

In a re-circulating cooling system, the cooling water carries waste heat removed from the steam turbine exhaust to the cooling tower, which rejects the heat directly to the atmosphere. Because of this direct path to the atmosphere, surrounding water bodies typically do not suffer adverse thermal effects. Some water is discharged from the cooling water system to maintain the concentration of chemicals in the cooling water below licensed limits. This water is often discharged to surrounding watercourses. In dry cooling systems, the waste heat is transferred directly to the atmosphere.

10. COGENERATION

10.1. Introduction

Cogeneration, also known as combined heat and power (cogeneration) or CHP, and total energy, is an efficient, clean, and reliable approach to generating power and thermal energy from a single fuel source. That is, cogeneration uses heat that is otherwise discarded from conventional power generation to produce thermal energy. This energy is used to provide cooling or heating for industrial facilities, district energy systems, and commercial buildings.

By recycling this waste heat, cogeneration systems achieve typical effective electric efficiencies of 50% to 70%-a dramatic improvement over the average 33% efficiency of conventional fossil-fuelled power plants. Cogenerations' higher efficiencies reduce air emissions of nitrous oxides, sulfur dioxide, mercury, particulate matter, and carbon dioxide, the leading greenhouse gas associated with climate change Cogeneration now produces almost 10% of our nation's electricity, saves its customers up to 40% on their energy expenses, and provides even greater savings to our environment.

Cogeneration, as previously described above, is also known as "combined heat and power" (CHP), cogen, district energy, total energy, and combined cycle, is the simultaneous production of heat (usually in the form of hot water and/or steam) and power, utilizing one primary fuel.

Cogeneration technology is not the latest industry buzz-word being touted as the solution to our nation's energy woes. Cogeneration is a proven technology that has been around for over 100 years. Our nation's first commercial power plant was a cogeneration plant that was designed and built by Thomas Edison in 1882 in New York. Primary fuels commonly used in cogeneration include natural gas, oil, diesel fuel, propane, coal, wood, wood-waste and bio-mass. These "primary" fuels are used to make electricity, a "secondary" fuel. This is why electricity, when compared on a btu to btu basis, is typically 3-5 times more expensive than primary fuels such as natural gas.

An example of a cogeneration process would be the automobile in which the primary fuel (gasoline) is burned in an internal combustion engine-this produces both mechanical and electrical energy (cogeneration). These combined energies, derived from the combustion process of the car's engine, operate the various systems of the automobile, including the drive-train or transmission (mechanical power), lights (electrical power), air conditioning (mechanical and electrical power), and heating of the car's interior when heat is required to keep the car's occupants warm. This heat, which is manufactured by the engine during the combustion process, was "captured" from the engine and then re-directed to the passenger compartment.

Due to competitive pressures to cut costs and reduce emissions of air pollutants and greenhouse gasses, owners and operators of industrial and commercial facilities are actively looking for ways to use energy more efficiently. One option is cogeneration, also known as combined heat and power (CHP). Cogeneration/CHP is the simultaneous production of electricity and useful heat from the same fuel or energy. Facilities with cogeneration systems use them to produce their own electricity, and use the unused excess (waste) heat for process steam, hot water heating, space heating, and other thermal needs. They may also use excess process heat to produce steam for electricity production. Cogeneration currently coexists with a regulated industry that is going through major structural changes that may limit or expand its application.

10.2. Cogeneration Technologies

A typical cogeneration system consists of an engine, steam turbine, or combustion turbine that drives an electrical generator is shown in figure 10.1. A waste heat exchanger recovers waste heat from the engine and/or exhaust gas to produce hot water or steam. Cogeneration produces a given amount of electric power and process heat with 10% to 30% less fuel than it takes to produce the electricity and process heat separately.

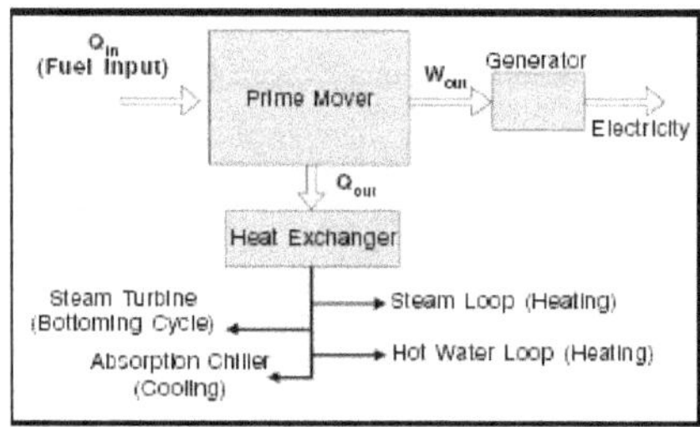

Fig. 10.1: Cogeneration Systems

There are two main types of cogeneration techniques: "Topping Cycle" plants, and "Bottoming Cycle" plants. A topping cycle plant generates electricity or mechanical power first. Facilities that generate electrical power may produce the electricity for their own use, and then sell any excess power to a utility. There are four types of topping cycle cogeneration systems. The first type burns fuel in a gas turbine or diesel engine to produce electrical or mechanical power. The exhaust provides process heat, or goes to a heat recovery boiler to create steam to drive a secondary steam turbine.

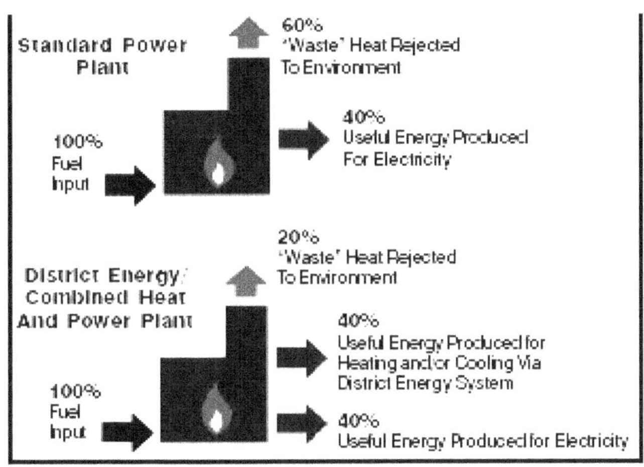

Fig. 10.2: The Advantage of Efficiency

The overall efficiency of the combined heat and power approach can exceed 80%. This is only possible where the generator is located at or near the facility and waste heat can serve the heating and cooling load. Even the highest efficiency electric utility plant barely exceeds 40% to 45%

This is a combined-cycle topping system. The second type of system burns fuel (any type) to produce high-pressure steam that then passes through a steam turbine to produce power. The exhaust provides low-pressure process steam as shown in figure 10.2. This is a steam-turbine topping system. A third type burns a fuel such as natural gas, diesel, wood, gasified coal, or landfill gas. The hot water from the engine jacket cooling system flows to a heat recovery boiler, where it is converted to process steam and hot water for space heating. The fourth type is a gas-turbine topping system. A natural gas turbine drives a generator. The exhaust gas goes to a heat recovery boiler that makes process steam and process heat. A topping cycle cogeneration plant always uses some additional fuel, beyond what is needed for manufacturing, so there is an operating cost associated with the power production.

Bottoming cycle plants are much less common than topping cycle plants. These plants exist in heavy industries such as glass or metals manufacturing where very high temperature furnaces are used. A waste heat recovery boiler recaptures waste heat from a manufacturing heating process. This waste heat is then used to produce steam that drives a steam turbine to produce electricity. Since fuel is burned first in the production process, no extra fuel is required to produce electricity.

An emerging technology that has cogeneration possibilities is the fuel cell. A fuel cell is a device that converts hydrogen to electricity without combustion. Heat is also produced. Most fuel cells use natural gas (composed mainly of methane) as the source of hydrogen. The first commercial availability of fuel cell technology was the phosphoric acid fuel cell, which has been on the market for a few years. There are about 50 installed and operating in the United States. Other fuel cell technologies (molten carbonate and solid oxide) are in early stages of development. Solid oxide fuel cells (SOFCs) may be potential source for cogeneration, due to the high temperature heat generated by their operation.

10.3. Technical Options for Cogeneration

Cogeneration technologies that have been widely commercialized include extraction/back pressure steam turbines, gas turbine with heat recovery boiler (with or without bottoming steam turbine) and reciprocating engines with heat recovery boiler.

10.4. Steam Turbine Cogeneration Systems

The two types of steam turbines most widely used are the backpressure and the extraction-condensing types (see Figure 10.3). The choice between backpressure turbine and extraction-condensing turbine depends mainly on the quantities of power and heat, quality of heat, and economic factors. The extraction points of steam from the turbine could be more than one, depending on the temperature levels of heat required by the processes. Another variation of the steam turbine topping cycle cogeneration system is the extraction-back pressure turbine that can be employed where the end-user needs thermal energy at two different temperature levels. The full-condensing steam turbines are usually incorporated at sites where heat rejected from the process is used to generate power.

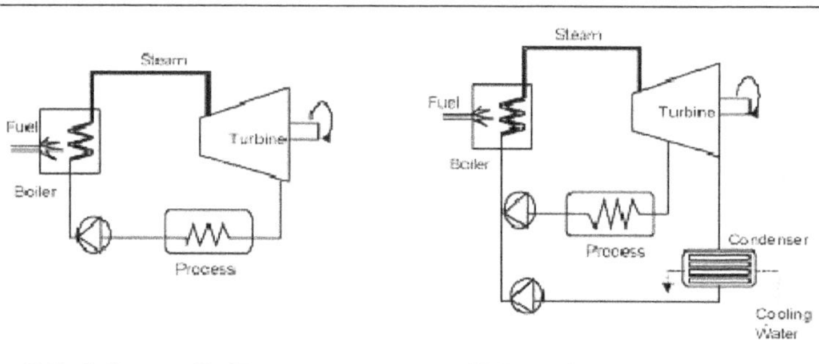

Fig. 10.3: Steam Turbines

The specific advantage of using steam turbines in comparison with the other prime movers is the option for using a wide variety of conventional as well as alternative fuels such as coal, natural gas, fuel oil and biomass. The power generation efficiency of the cycle may be sacrificed to some extent in order to optimize heat supply. In backpressure cogeneration plants, there is no need for large cooling towers. Steam turbines are mostly used where the demand for electricity is greater than one MW up to a few hundreds of MW. Due to the system inertia, their operation is not suitable for sites with intermittent energy demand.

10.5. Gas Turbine Cogeneration Systems

Gas turbine cogeneration systems can produce all or a part of the energy requirement of the site, and the energy released at high temperature in the exhaust stack can be recovered for various heating and cooling applications (see Figure 10.4). Though natural gas is most commonly used, other fuels such as light fuel oil or diesel can also be employed. The typical range of gas turbines varies from a fraction of a MW to around 100 MW. Gas turbine cogeneration has probably experienced the most rapid development in the recent years due to the greater availability of natural gas, rapid progress in the technology, significant reduction in installation costs, and better environmental performance. Furthermore, the gestation period for developing a project is shorter and the equipment can be delivered in a modular manner. Gas turbine has a short start-up time and provides the flexibility of intermittent operation. Though it has a low heat to power conversion efficiency, more heat can be recovered at higher temperatures. If the heat output is less than that required by the user, it is possible to have supplementary natural gas firing by mixing additional fuel to the oxygen-rich exhaust gas to boost the thermal output more efficiently.

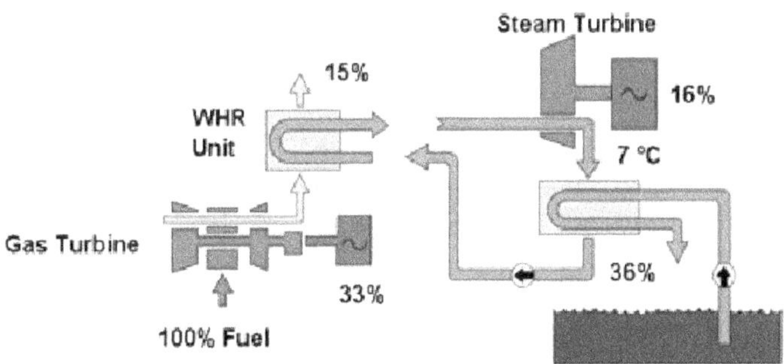

Fig. 10.4: Gas Turbine

On the other hand, if more power is required at the site, it is possible to adopt a combined cycle that is a combination of gas turbine and steam turbine cogeneration. Steam generated from the exhaust gas of the gas turbine is passed through a backpressure or extraction-condensing steam turbine to generate additional power. The exhaust or the extracted steam from the steam turbine provides the required thermal energy.

10.6. Reciprocating Engine Cogeneration Systems

Also known as internal combustion (I. C.) engines, these cogeneration systems have high power generation efficiencies in comparison with other prime movers. There are two sources of heat for recovery: exhaust gas at high temperature and engine jacket cooling water system at low temperature (see Figure 10.5). As heat recovery can be quite efficient for smaller systems, these systems are more popular with smaller energy consuming facilities, particularly those having a greater need for electricity than thermal energy and where the quality of heat required is not high, e.g. low pressure steam or hot water.

Though diesel has been the most common fuel in the past, the prime movers can also operate with heavy fuel oil or natural gas. These machines are ideal for intermittent operation and their performance is not as sensitive to the changes in ambient temperatures as the gas turbines. Though the initial investment on these machines is low, their operating and maintenance costs are high due to high wear and tear.

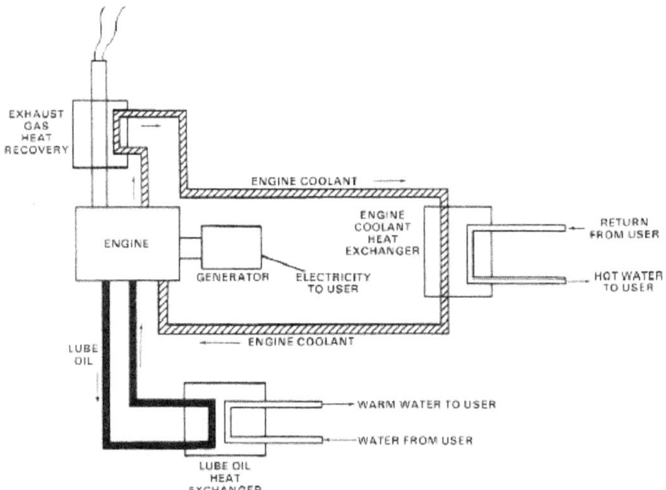

Fig. 10.5: Reciprocating Engine Cogeneration Systems

10.7. The Advantages of Combined Heat and Power Production

In the back-pressure CHP plant all steam is condensed at a relatively high coolant temperature, thus reducing the electric output to approx. 36% compared with a condensing plant of efficiency 45%, however, an increase In the overall efficiency (up to 92%) is achieved by a heat output of 56%. The back-pressure CHP plant has a fixed ratio. C_m of power to heat output as shown in the example below:

$$C_m = 36\%/56\% = 0.64$$

The advantage of CHP production Is illustrated by comparing two large plants without and with CHP: a condensing plant (45%) and a back-pressure plant (36% and 56%):

Reduced power production at the back- pressure plant(calculated as units of energy produced per 100 units of energy in the fuel): 45 - 36 = 9 units

Achieved heat production: 56 units

The change from a condensing plant to a back-pressure plant with the same fuel input (100 units) thus means that 9 units of electricity have to be produced elsewhere. e.g. at another condensing plant. If this is the case "energy efficiency "for the heat production can be calculated comparing the heat achieved at the back-pressure plant with the additional fuel used at the condensing plant:

Additional fuel used for producing 9 units of power at a condensing plant:

$$9 \text{ units}/0.45 = 20 \text{ units}$$

"Energy efficiency" of the achieved heat production:

$$56 \times 100/20 = 280\%$$

This means that in order to produce 1 unit of heating, only fuel corresponding to about one-third of the unit is needed. This calculation is relevant only in full back-pressure operation with an overall efficiency of more than 90%. Modern extraction CHP plants operate between the two regimes depending on the heat consumption relative to the electricity production. In such a case the annual efficiency of a CHP plant Is typically 70-75% and the "energy efficiency" of the heat production achieved is typically 200%.

It follows from the calculations that CHP is important to make rational use of energy. Comparing with a condensing power station a CHP plant approximately doubles the utilization of the energy content of the fuel.

Another rule of thumb Is the following: CHP saves about 30% fuel as compared with separate production of electricity and heat.

10.8. Large-Scale Plants

a) Condensing Power Plants

Water is pumped into the boiler at high pressure and then vaporised. Generally, fossil fuels supply the heat. From the boiler the steam Is led to the turbine normally at a pressure of about 240 bars and temperature of 560°C when the power station is a large one.

In the turbine, the thermal energy of the steam is converted into mechanical energy to rotate a generator shaft. Approximate steam conditions following the energy transfer are0.04 bar and 40°C. Steam condensation takes place in a condenser, which is cooled with large quantities of water. In Denmark mainly seawater is used for this purpose, heated 8-10°C In the condenser. The condensed steam (condensate) is then pumped back into the boiler.

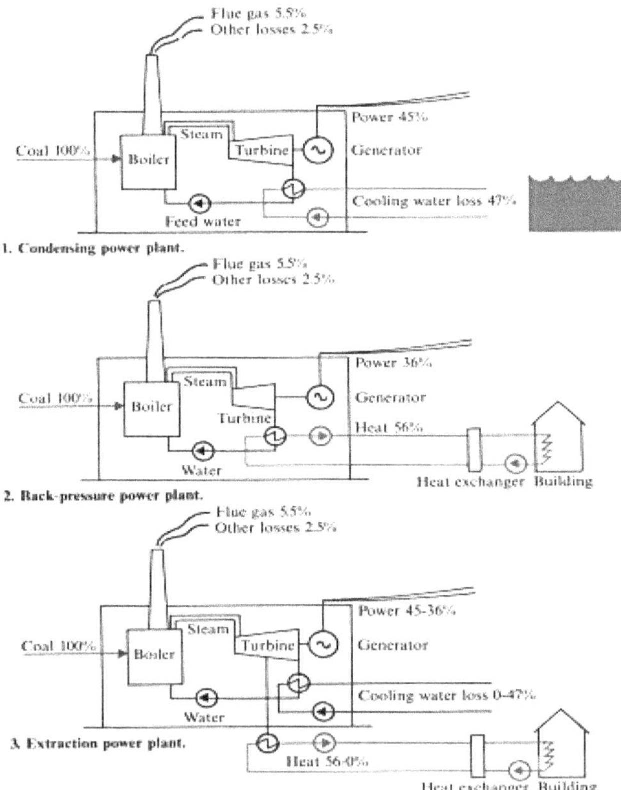

Fig. 10.6: Large-Scale Plants

b) Back-pressure CHP Plants

As indicated in figure 10.6 (1) about 47% of the fired energy is rejected with the cooling water in a condensing power plant. Some of this heat can be utilized In a CHP generation process by raising the temperature of the supply water of a DH system. In a conventional power plant the steam Is condensed at about 40^0 C a temperature too low for supplying DH. In a CHP plant It Is therefore necessary to condense the steam in the condenser under pressure and temperature conditions that enable DH water to reach 85-120°C.

In the back-pressure plant the ratio of electricity to heat production Is fixed. This Is a disadvantage that can be somewhat alleviated by using heat accumulators.

c) Extraction CHP Plants

This type of plant is characterized by a variable ratio of power to heat generation (fig. 10.6. (3)). The DH water Is Indirectly heated by steam extracted from the turbine. The extraction arrangement at the turbine enables the DH water to be heated to l00-120°C. The greater the steam extraction for hot water production, the smaller the power generation. This means that when more hot DH water is produced, less heat energy is lost at the con denser.

In the case of large CHP units (125 to 500 MW_e). the production of hot DH water takes place almost exclusively at extraction power plants because of their great flexibility.

d) Combined Cycle CHP Plants

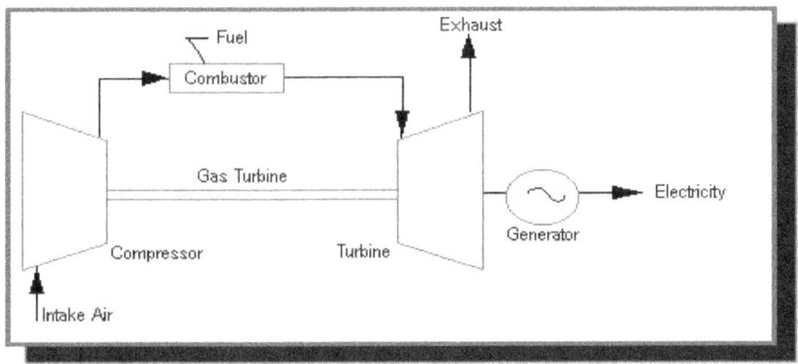

Fig. 10.7: Combined Cycle CHP Plants

The hot flue gas from the gas turbine is utilized in the boiler (see fig. 10.7) to produce steam for the steam turbine. It is necessary to use either natural gas or fuel oil for the gas turbine, whereas it is possible to use either gas or coal as supplementary fuels for the steam boner.

The steam turbine of a small-scale combined cycle plant is normally a back-pressure unit, whereas a large-scale plant normally has an extraction turbine.

As an example, a 350 MW extraction plant will have the following main data: Power, max. 350 MW

Efficiency, condensing operation 55%

Power back-pressure operation 315 MW

Heat, back-pressure operation 250 MJ/s

C_m 1.26

Efficiency, back-pressure operation 88%

10.9. Small-Scale CHP Plants

Small-scale CHP plants can be placed close to the heat consumers. The costs of the transmission pipe system will thus be relatively low compared to the case of large plants. Some examples of such types are shown below.

a) Combustion-type Engines

Combustion-type engines are now available in sizes up to about 47 MW and down to a few kW Electrical efficiencies range from about 30% for the smallest and 35-40% for ignition lean-burn gas engines. For large diesel engines the electrical efficiencies can be as high as 50%. An example of a combustion engine with turbo-charged for natural gas equipped with a waste heat boiler is shown in figure 10.8. Due to the utilization of heat it is possible to attain a total efficiency above 85% and CM values ranging from 0.85 to 1.0.

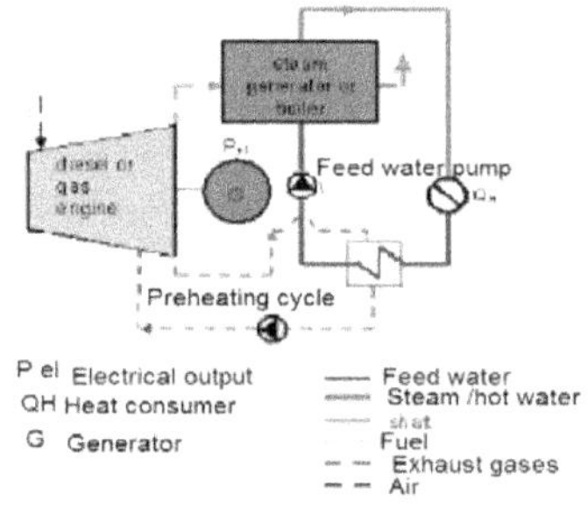

Fig. 10.8: Small-Scale CHP Plants

b) Gas Turbines

A CHP plant based on a gas-fired gas turbine is shown in fig. 10.9. Gas turbines range in size from 200 kW to 220 MW. Their electrical efficiencies are about 20% for the smallest units to 40% for the largest. Figure 10.9 shows an example of a medium-size local CHP plant equipped with a waste heat recovery boiler. Total efficiencies above 85% are possible and CM values are In the range of approx. 0.5 -0.6.

c) Combined-cycle CHP Plants

The principle of a combined-cycle CHP plant has been described earlier (fig. 10.7). The minimum size will be approx.10 MW power. The small-scale plants are normally back-pressure units with a CM value of approx. 1.0.

d) Biomass or Coal-fired CHP Plants

These plants use wood chips, straw, waste or coal to fuel a steam boiler supplying a back-pressure steam turbine (fig. 10.6(2)). Due to low steam data and no reheating of steam, the CM values are rather low for this type of CHP plant, viz. 0.3-0.5. However, biofuels have the environmental advantage of being close to CO_2 neutral.

In the field of biomass extensive research and development efforts are being made to develop methods for gasifying the biomass in order to use combined cycle plants or combustion engines and attain a higher CM value.

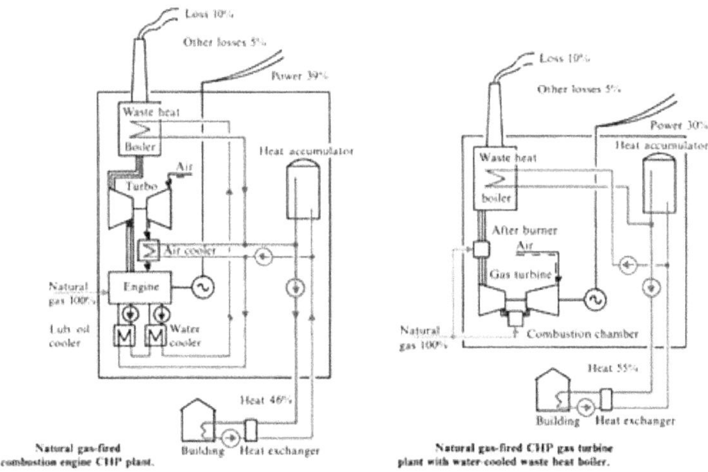

Fig. 10.9: A CHP Plant based on a Gas-fired Gas Turbine

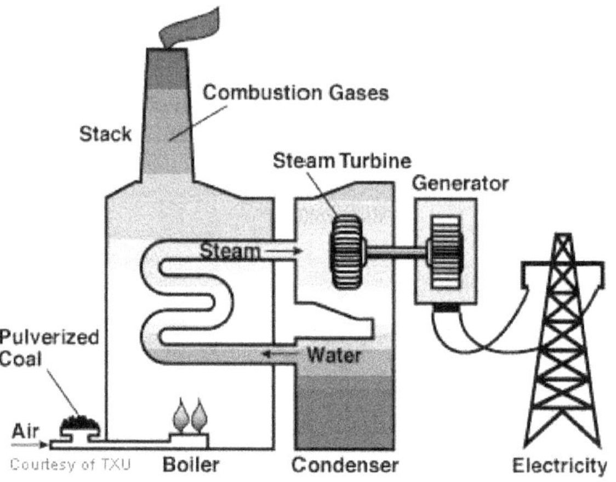

Fig. 10.10: System Design and Optimization of a CHP System

When planning the implementation of a CHP system, whether it be a renovation or the establishment of an entirely new system, it is important to assess and optimise the system design.

It is particularly important to choose the correct levels of supply and return temperatures. For the previously mentioned CHP plants it is assumed that the supply temperature will be 100°Cand the return temperature 50°C.

Lowering the level of district heating temperatures will raise the CM value and vice versa.

The optimization of a CHP plant should include the transport costs for heat. Likewise the possibility of reducing the costs of production and transport of heat by means of heat accumulators should be included in the optimization.

Future progress in savings and heat insulation of buildings as well as new connections to customers and new supply areas will also influence total system design. Operating the system at a lower temperature level will typically result in increased transport charges but overall decreased production costs.

Increasing the CM value of a system will usually prove to be economically advantageous. It is possible to improve this CM value by changing either the production technology or the pipeline network. Sometimes improvements may be made at a considerably lower cost by altering the production technology rather than the pipeline network.

Typical guidelines of optimal supply-pipe temperatures are shown in figure 10.10.

10.10. The Development Trend of CHP Plants

Higher Electric Efficiency and Higher CM Value

a) Combustion Engines

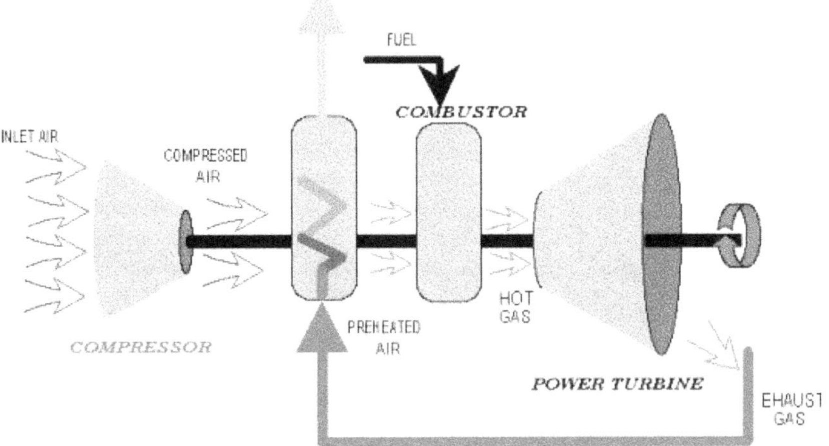

Fig. 10.11: Gas Turbine with Regeneration

In recent years, gas engines based on natural gas have improved vigorously towards higher electric efficiencies and thereby higher CM values as shown in the figure 10.11. For lean-burn engines of sizes 1-2 MW electric efficiencies are approaching 40%. The trend of the development has been to optimize the control of the combustion process by using a combustion prechamber and increasing the excess air factor. *, to 2.0 for lean-burn engines. By so doing, the mean effective pressure in the combustion process is increased and the NO_x formation kept at a low level.

b) Back-pressure CHP Plants

High-temperature-resistant ferritic steels used in such places as boiler super heaters and live steam pipes offer high flexibility at elevated steam parameters in contrast to austenitic steels. They play a key role in improving the ability of thermal power stations to attain higher efficiencies, lower pollution levels, lower fuel consumption, and lower manufacturing costs. Based on present development it can be assumed that with the improved ferritic steels containing 9-10% CrMoVNb. The present limit of the steam admission temperature of 560°C can be raised to approximately 600°C. The modified 9% Cr- Steel (P91/T91) constitutes a good example of this steel category. By raising the steam temperature from 560 to 600°C, the electric efficiency of a condensing plant can be increased 2 percentage points. For a back-pressure plant, the electric efficiency will increase by approx. 3 percentage points corresponding to an improvement of about 5 percentage points in the CM value. As can be seen, the elevated steam data may be even more relevant for a back-pressure plant than for a condensing unit.

c) Gas Turbines

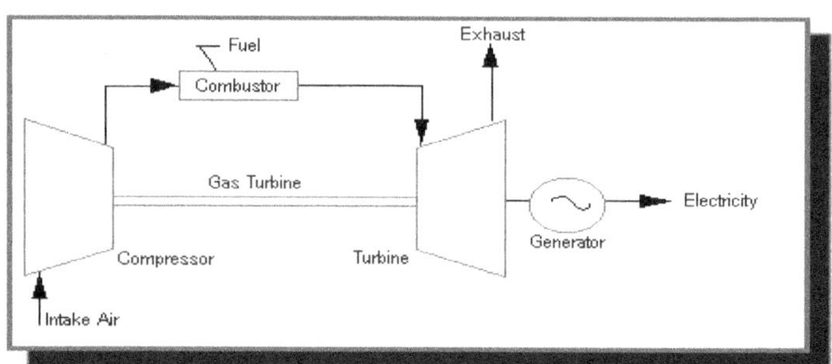

Fig. 10.12: Gas Turbines

The electric efficiency, and thus the CM value, is usually increased by raising the flue gas temperature at the inlet for the first row of blades as much as possible as shown in the figure 10.12. e.g. to 1100ºC. This requires improved materials in the combustion chamber and turbine blades as well as an advanced cooling technique for the components. Besides, certain parts of the blades may be coated. The combustion process itself is controlled in order to obtain the lowest possible NO, formation. Today this is done by means of a special design of the combustion chamber called dry low-NO, type, while the design used up till now has been water or steam injection. Today the air volume can be regulated down to 75-80% by means of stationary blade shifting. This enables the electric efficiency to be increased at part load.

10.11. Cogeneration Benefits

Cogeneration offers energy, environmental, and economic benefits, including:

Saving Money

By improving efficiency, cogeneration systems can reduce fuel costs associated with providing heat and electricity to a facility.

Improving Power Reliability

Cogeneration systems are located at the point of energy use. They provide high-quality and reliable power and heat locally to the energy user, and they also help reduce congestion on the electric grid by removing or reducing load. In this way, cogeneration systems effectively assist or support the electric grid, providing enhanced reliability in electricity transmission and distribution.

Reducing Environmental Impact

Because of its improved efficiency in fuel conversion, cogeneration reduces the amount of fuel burned for a given energy output and reduces the corresponding emissions of pollutants and greenhouse gases.

Conserving Limited Resources of Fossil Fuels

Because cogeneration requires less fuel for a given energy output, the use of cogeneration reduces the demand on our limited natural resources-including coal, natural gas, and oil-and improves our nation's energy security.

10.12. Where can cogeneration are used?

Cogeneration installations are most likely to be economically viable at locations where the following characteristics exist:

- Coincident demand for electricity and thermal energy (i.e., steam, heating, or cooling) during most of the year.
- Access to fuels, including natural gas, biomass, and/or by-product fuels.

The Following are Typical Markets for Cogeneration

Energy-intensive industries, including the chemical, refining, forest products, food, and pharmaceutical sectors.

District energy systems that distribute heat or chilled water to a network of buildings. Such systems show the greatest promise in downtown areas, industrial parks, college campuses, military bases, and other large institutional facilities.

High power reliability/quality applications, such as Internet or telecommunications data centres requiring high-quality, reliable power and substantial cooling capacity.

Institutional markets, including hospitals, hotels, and convention centers where large year-round demands exist for electricity, heating, and cooling.

Abandoned industrial sites, or brown fields, where cogeneration-based systems can provide the energy infrastructure for "power parks," facilitating economic redevelopment of underutilized properties. Commercial buildings-as building-scale cogeneration technologies become better integrated and increasingly cost-effective, this market offers large potential for new applications.

A Small Sample of Successful Businesses Now Using Cogeneration Includes

Agriculture, apartment buildings, auto/car dealerships, casinos, cold storage facilities, communications sites, convenience stores, credit card processing facilities, customer service centres, dairies, fabrication plants, feed yards, foundries, golf courses, government buildings, commercial greenhouses and nurseries, grocery stores, hospitals, hotels, ice skating rinks, industrial parks, ISP's, landfills, laundries/Laundromats, malls, manufacturing plants, military bases and installations, motels, nursing homes, oil & gas leases, office buildings, paper & pulp, parking garages, printing companies, processing plants, radio stations, resorts, restaurants, retail stores, retirement homes, schools, server farms, shopping centres, sports complexes, steel manufacturing, supermarkets, television stations, universities, warehouses, waste treatment facilities, wineries

11. Operational Problems and Emergency Cases in Thermal Power Plant

11.1. Introduction

For good operator and the designer of the power station, it is very essential to understand and analyze the operational problems and the emergencies of common nature which are likely to arise in the main equipment viz. Boiler, turbine and generator and other auxiliaries during operation.

A thorough understanding and analysis of all types of malfunction, their causes, plant behaviour and operator action to tackle them is also very important for an operator. In operation of large thermal sets a malfunction occurring in one equipment, causes abnormal behaviour in other connected equipment also. The success of operation of a set lies in being forewarned about such possibility, and being prepared to tackle them as they arise. Such systematic action not only leads to safe and quickest elimination of malfunction, but also results in maximizing generation from the set, avoiding mishap and costly breakdowns.

Because of higher parameters of pressure, temperature speed and also due to a large number of auxiliaries, the chances of emergency situations are more in thermal power plant than in hydro or diesel power plant. It is important to note that a better maintained plant will lead to lesser operational problems and similarly, good operation will require lesser maintenance. Thus both maintenance and operation are very important aspects. Sometimes negligence on the part of operation personals may not have any immediate noticeable problem, but in the longer run it will be injurious to the plant.

The operator has to ensure maintenance of critical parameters at desired values and run the power plant most efficiently without allowing dangerous conditions leading to risk equipment reach. In emergency situation they have to act most intelligently keeping a strict watch on health of plant. In order that operation personnel may act correctly and efficiently in cases of emergency, they should have a thorough knowledge of the plant, process parameters, interlocks and protection. They should have been thoroughly trained on simulators to act in emergencies.

Though interlocks and instrumentation, protections, and auto controls are designed to achieve all these factors automatically and minimize dangers arising due to human error and negligence but nevertheless human element has to play its role. If any of the controls or protections fail, the operator should be able to take corrective measures.

Broadly speaking the emergencies may be of nature which requires:

- Shutting down the unit due to problems in boiler.
- Shutting down the unit quickly and stopping the turbine with the help of vacuum break so as to bring turbine to standstill in least possible time.
- Shutting down the turbine without applying vacuum brake and let boiler run through HP/LP by pass.
- Protections requiring load reduction on turbine.
- Protections requiring partial load operation of the unit at 50 – 60%of rated load.
- Protections of auxiliaries of boiler and turbine.

House load operation, i.e. operating boiler and turbine to feed its own auxiliaries. Such a condition arises when generator breaker is tripped due to external fault in transmission lines. In many cases immediate stoppage of the plant may not be essential, and emergency can be taken care of by taking certain remedial measures. It should be appreciated that frequent trappings lead to high stress strain on plant and availability of plant is reduced.

11.2. Emergency Cases Requiring Shut Down of Boiler

1. Very High drum level
2. fall in drum level
3. pressure rise in boiler furnace
4. low furnace pressure
5. flame failure
6. Both ID and FD trip.

11.2.1. Very High Drum Level

This is harmful for turbine because high level of drum may result in carryover of water to super-heater and from there to main steam line and in extreme cases to turbine as well. This will result in server hammering all over and damage to the equipment .It is also a possible source of salt deposition in turbine blades. At high drum level, superheated steam temperature will also drop sharply. Such a condition could occur due to control of boiler feed valves, failure of auto controller ; failure or inaccuracy of gauges, thus misleading the operator ; leakage through other valves on feed line, sudden increase of load; sudden increase in firing rate.

Before occurrence of such a condition, operator should check performance of gauge; take over feed control valves on remote manual control and decrease the flow of water, try load reduction for controlling water level, if high water level is due to upward load surge; check valves in by pass lines(if they are passing). During the entire operation of high level, keep a stricter control on steam temperature; open emergency blow-down valves. If all the measures have failed, trip the turbine and open drain before turbine on the main steam line.

11.2.2. Fall in Drum Level

The causes for fall in drum level are: Fall of pressure in feed line; fault of auto control valve on feed line; tripping of one of the working feed pumps; faults in level gauges; sudden reduction of load; sudden tripping of one or more mills, oil burners; leakages in the boiler drains, valves and fittings; damages in water heating surfaces, such as economizer water walls, super-heater etc; inadvertent opening of drum emergency drain valve. On getting drum level alarm, operator should check the steam consumption and water flow and find out the reason for differences, start standby pump if it has not stated automatically; blow down the gauges in order to check their performance, take feed control on manual and increase feed-water flow (it is important to note that too high discharge from working pump is not good for pump).

Check for leakage(leakage in water wall and super heater will be accompanied by raise in pressure in furnace and hissing noise, leakages are established and it is difficult to maintain level the unit should be tripped. Reduce the load, if it has been decided to continue the operation.

11.2.3. Pressure Rise in Boiler Furnace

Any abnormal rise in pressure will result in back firing or explosion in the boiler or along the flue path.

1. Furnace pressures my increase due to tripping of one of the two ID fan of regulating vanes or Fans/closing of dampers on the flue gas side.
2. Damage to heating surface.
3. Un stable coal flame.
4. Low wind box pressure, improper air distribution at elevation, too much or too low fuel air, sudden starting of mill, loss of ignition energy.
5. Unequal burner tilt at corners.
6. Burner tilt in extremes (up or down), gradually built fouling in air heater or stoppage or air heater.
7. Furnace water seal broken
8. Manholes in electrostatic pressure or elsewhere in the flue gas track suddenly open.
9. Gas explosions in flu gas path
10. ID Fan/PA fan air out of control(excess air)

High furnace pressure may lead to possibilities of minor or major furnace explosion, make combustion unstable, flue gas escape from manhole, peepholes etc; boiler pressure drops; igniters may trip out causing associated problems. Under such a condition:

1. Operator should check by draft readings whether and damper is closed .
2. Check vane control mechanisms of fans, motors current values, correct possible malfunction if an ID fan has tripped, reduce load, stabilise on one ID fan and then restart other fan.
3. Check furnace seal, which may get broken by sudden slag fall, or by low or interrupted water supply (ensure that water flows through overflow overflow drains).

If sudden starting of the mill causes furnace pressurization wait for a few minutes, as it may get normalized.

11.2.4. Low Furnace Pressure

Too low furnace pressure may blow out flame from working igniters or even oil guns, and may cause unstable flame conditions.

1. Furnace pressure may be low due to ID fan auto control failure.
2. ID fan vane control mechanism failure causing vanes to open wide.
3. Sudden load throw off, sudden decrease in air input or tripping of one FD fan.

Under such condition

a. Operator should switch ID fan control to manual and bring normal parameters.
b. Check ID fan vane control mechanism, check air flow, and restart FD fan.

11.2.5. Fame Failure

It may occur due to

1. Closing of trip of heavy oil/warm up oil/igniters when fuel oil is being used as a support fuel or when being used as a main fuel.
2. Sudden decrease of mill feeder speeds to minimum
3. loss of ignition support energy
4. Sudden closure of auxiliary, fuel air dampers to burner nozzle.
5. too low or too high wind box pressure causing unstable combustion
6. water wall tube failure (jet extinguishes flame)
7. Malopertion of soot blower nozzles, causing failure of scanner air fans, slag or soot on scanner lenses etc., burner tilt negative extreme due to air failure to power cylinders.
8. Sudden coal hangs ups in any one or two mills.

Prior to failure

1. Operator will notice fluctuation in furnace draft

2. flickering of flame scanners

3. Tripping or PA fans or low PA pressure.

4. Coal air temperature after mill high

5. Oil burner/ignitor failure.

11.2.6. Both ID and FD Fans Trip

Such a condition could occur due to

1. 6.6KV supply failure.

2. 0.4KV supply failure

3. Cooling water to motors failure, or bearing temperature getting too high due to trouble in lub oil system etc.

Under above condition and all other condition leading to boiler trip

1. Operator should start preparation for hot rolling and ensure that vacuum system is OK.

2. Start the starting oil pump, reset turbine trip supervisory relay.

3. watch turbine expansion

4. Check fed pump, maintain drum water level.

5. Complete purge

6. Introduce furnace probe, light up boiler ,bring to service LP/HP bypass

7. Mach boiler outlet/turbine metal/steam temperature.

8. Roll turbine (after closing LP bypass) synchronies generator to grid, stop starting oil pump.

9. Quickly start PA fan/Milling system and load generator, to maintain steam temperatures.

10. Watch differential expansion and maintain drum level.

11.3. Shut Down of Unit Requiring Use of Vacuum Breaker

As far as possible, consistent with the safety of generator and turbine, use of vacuum brake should be avoided as on breaking the vacuum, rotor is subject to sudden braking and may cause local heating. However under the following abnormal conditions, vacuum needs to be broken by opening the value connecting condenser to atmosphere.

1. Abrupt increase in axial shift up to critical limit.

2. Rise in temperature of oil.

3. Emission of sparks from rear seals for turbine.

4. Sudden drop in turbine oil level and pressure.

11.4. Shut Down of Turbine without Use as Vacuum Breaker

The turbine will stop on the appearance of following emergencies:

1. Sudden fall of vacuum unto lower critical limit

2. Tripping of both circulating water pumps.

3. HP heater level very high

4. All boiler feed pumps tripped.

5. Very low main steam temperature

6. Tripping of generator due to its electrical protections.

7. Closing of both emergency safety valves.

8. Fire in oil system.

11.5. Other Emergency in Boiler

1. Water wall tube failure.

2. Reheater/super heater/economizer tube failure.

3. High reheat steam temperature.

4. Steam pressure rise

5. Ignition of un burnt particles of coal in boiler flues.

6. plugged air heater

7. Fire in air heater

8. Slag in water walls

9. Slag in super heater

10. Boiler water conductivity high

11. Impure steam

12. High smoke density

13. Particulate emission high

14. Boiler flue gas temperature low

11.5.1. Water Wall Tube Failure

It could occur due to

1. starve water wells

2. sustained flame impingement on water wall tubes

3. blocked tube, eroded tube, pitted tube, salt deposition

4. soot blower steam impingement

5. Circulation affected due to open low point drains.

Symptoms of bursting of water walls are

1. "Hissing" noise created by steam leakage from the boiler, noticeable a few hours   in advance before situation becoming serious.

2. Unstable flame, fluctuating draft

3. Flame failure

4. Marked higher feed flow for a given steam generation, overloading of ID fans,

5. Pressure and steam flow

Under such a condition, operator should start immediately load reduction, trip out unit before damage becomes serious. Quicker the shutdown, possibilities of expensive damage can be avoided. Boiler feed should be discontinued and ID fans allowed to run.

11.5.2. Reheater/Super Heater/Economiser Tube Failure

These could occur in reheater/superheater due to

1. Sustained high metal temperatures due to water wall staging or rapid rise of load on boiler at the time of hot restarts (creep failure).

2. Erosions of tube due to high erosive as him flue gases and dislocated tubes.

3. Blocked tubes, erosion due to shoot blowing, salt deposition due to high water level in drum, poor quality spray water, and poor quality water during hydraulic tests.

4. Inadequate flow due to open super heater drains.

Symptoms of bursting of superheater/reheater tubes could be:

1. "Hissing noise" noticeable hours before other symptoms

2. Unexplained drop of flue gas temperature in the affected region

3. Overloading of ID fan /increased draft losses, higher feed water consumption

4. Superheat/reheat temperature drop

5. High superheater/reheater metal temperature

6. Pitted/corroded tube thinned out at the bends.

7. Ash erosion due to soot blower nozzles.

Such a condition can be judged by increased draft loss/ID fan loading, drop in flue gas temperature after economizer: noise in the economizer zone; ash solidification in economizer hopper; water leakages from economizer hoppers; and high feed water consumption. As soon as leak is detected operator should start to reduce load and trip out boiler at the earlier possible time. He should try to locate leak through manholes before boiler is depressurized.

11.5.3. High Reheat Steam Temperature

These lead to high metal temperature and rise in positive turbine expansion.

Under such a condition, operator should check the performance of

1. Auto injection controller, if found incorrect, resort to remote manual control, check the flow of water in to the attemperator according to the ridings of the flow maters and steam temperature prior to and after the injection.
2. Control combustion (air distribution etc.) and reduce firing rate, check water temperature and bring in heaters.
3. Reduce air flow
4. Initiate water wall soot blowing, change over to lower mills, in possible.

If temperature on higher side persists for 3 minutes and the unit has not tripped, shut down the unit.

11.5.4. Steam Pressure Rise

This will result in blowing of safety valves, fluctuation of drum level, and shooting up of auxiliary steam pressure. The causes for rise in steam pressure could be partial decrease of load; full throw off-of load; wrong performance of combustion controller; higher firing rate; uncontrolled coal flow through mills. Symptoms of full/partial throw off-of load are:

1. Abrupt rise of steam pressure in the drum and stem mains, decrease in stem flow.
2. Abrupt change in drum level, fall in MW reading increase in turbine speed, fall of steam pressure in turbine first stage pressure.

Under such a situation operator should switch over steam pressure controller to manual and reduce coal feed. If pressure reaches to a pre-determined limit, cut off some of the mills. Drum level and superheat temperature control should be under close observation and regulated to maintain within limits. If safety valves do not blow then either valve on steam blow down line should be opened or safety valves opened from control room through remote operation. Tripping of oil burners should be avoided as furnace stability may be disturbed due to sharp decrease in coal firing rate which may have to be done to control steam temperature. As soon as pressure starts dropping firing rate may be normalized.

11.5.5. Ignition of Unburnt Particles of Coal in Boiler Flues

Such a condition can be judged from abrupt from abrupt increase of temperature in the flue over-heating of gas duct lining, escape of smoke or flame from the manholes, increase of pressure in boiler furnace and abrupt increase in ID fan motor current.

Under such a condition:

1. Furnace ID fans, pulverized fuel may be shut down.
2. Dampers on air and gas ducts closed tightly
3. Steam line to convective shaft may be opened to fill the place

11.5.6. Plugged Air Heater

This condition could occur due to defective oil combustion leading to soot, and infrequent operation of AH soot blowers. This is detected by draft of FD/PA/ID fans for same output, fall in hot air temperature, and rise in flue gas outlet temperature. Under such a condition operator should carry out soot blowing, reduced load on generator; restrict flue gas outlet temperature to ID fan capabilities by reduction in load.

11.5.7. Fire in Air Heater

This should occur due to defective oil combustion, leading to heavy soot/unburnt oil particles in air heater surfaces and in frequent/ineffective operation of soot blowers. This can be detected by infrared detectors or by checking air heater air temperature and flue gas temperature rising sharply. Under such a condition:

1. Operator should trip unit.
2. stop all fans
3. close all dampers
4. flood air heater with eater
5. Continue air heater rotation, at least with air motor till it is possible.

11.5.8. Slag in Water Walls

Slag accumulates on water walls continuously during coal firing, but abnormal operating conditions accelerate the deposits. Slagging is accelerate due to change of coal properties, improper air distribution between corners, too fire or too coarse a coal, low excess air excessive negative burners tilt, high firing rate caused by low feed water temperature, improper fuel air distribution etc.

11.5.9. Slag in Super Heaters

Slag or ash accumulation in super heaters causes low superheat temperature and high boiler exit flue gas temperature. This can be detected by pall of superheat/ reheat temperature; tilting of burner tilt upwards to correct the temperature; and increased super heater draft loss.

11.5.10. Boiler Water Conductivity High

Too high conductivity leads to scale formation in tubes if sustained over a period of time. This may occur due to low a blow down; condensate contaminated and poor quality of makeup water. If increased blow is unable to keep up with rise of conductivity, load may be reduced gradually unit shutdown.

11.5.11. Impure Steam

Steam is considered impure; if there is high silica in steam and phosphate carry over exists. Solids carry over in steam lead to deposits on turbine blades resulting in increase of first stage pressure and gradual reduction in turbine output. To obviate this situation, operator should check silica in boiler water which if high, blow down on boiler should should be increased boiler pressure/turbine load should be reduced to confirm to curve " silica in boiler water/ boiler pressure" taking into account solubility of silica in steam at elevated pressures. Drum level should be maintained at low level. If phosphate concentration in drum is more that 5 ppm, phosphate dosing should be stopped to stabilize at lower permissive value. If steam quality cannot be improved unit should be shut down.

11.5.12. High Smoke Density

It may be due to improper oil combustion, which in turn may be due to low or high atomizing medium differential pressure, inadequate air; or choked oil guns. It can be sensed black smoke emission from chimney, unstable combustion, possible fall of boiler pressure or temperature, possible carryover of unburnt oil particles to air heaters and emission from chimney. Under this condition, operator should check fuel oil pressures, temperature, atomizing medium pressures as per requirements, increase wind box pressure, air flow, and check opening of auxiliary air dampers/closing of fuel air dampers; and scavenge burners.

11.5.13. Particulate Emission High

High ash discharge through chimney, possible due to failure of EP or ash handling system may lead to premature failure of ID fans. In cities it is a violation of civic codes and where such legislation exists, is a violation punishable by law. It may occur due to failure of ash handling system leading to ash build up, or under this condition, operator should isolate gas passes one by one to eliminate possible defective pass, and reduce lead to contain particulate emission.

11.5.14. Boiler Flue Gas Temperature Low

Low flue gas temperature at outlet, if allowed over a period of time, may cause low temperature corrosion of air heaters. Low flue gas temperature can be a symptom of some other malfunctions which can have serious repercussion. Under such a condition operator should charge steam coil air heater at lower loads and ensure that their drains are functioning normally. At high loads, it should be ensured that all LP & HP heaters are charged and feed water temperature is not less than specified; air heater seal leakages are within permissible values.

11.6. Emergencies in Turbine

1. High differential expansion.
2. First stage pressure high
3. Turbine vibration high
4. Turbine eccentricity high
5. water hammer in main steam line
6. Condenser tube failure.

11.6.1. High Differential Expansion

As the rate of expansion or contraction or rotor under most of the condition is different to that of turbine stator, the rotor relative stator might elongated or contract. The tolerances for relative expansion are higher than relative contraction. If safe limits have been crossed, there might be appearance of vibrations and metallic sound and high temperature of oil from drains.

1. The causes of relative elongation of rotor: i.e. positive differential expansion
2. The cause of relative shrinkage of turbine rotor; i.e. negative differential expansion

11.6.2. First Stage Pressure High

This may occur due to:

1. salting in the turbine blades, closure of one of the interceptor valves or control valves of intermediate pressure cylinder
2. high load when HP/LP heaters are not in service
3. operating with low steam temperature on high loads or with full open servo-motor control valves

1. Overloading of Turbines

As a result of high first stage pressure, axial shift may increase thrust load on turbine may increase with possible premature wear of thrust pads; extraction pressure at selected points may increase and blades may be over stained. Under such a condition the operator should:

1. Reduce load immediately to restrict first stage pressure.
2. Check that interceptor valves and controls on IP are open
3. If salting in turbine is suspected, turbine should be steam washed during next cold start up.

Operator should also maintain steam purity, set boiler drum water level normal and maintain boiler steam parameters consistent with turbine load as per recommendations.

11.6.3. Turbine Vibration High

The possible causes of high vibration could be steam inlet pressure and temperature sudden change, Lube oil temperature and pressure too high or too low after the coil coolers, destroying oil film under the bearings, gland steam temperature low, temperature difference between top and bottom cylinder exceed permissible value (50 C) during start ups or during steam parameter and load changes; high turbine differential expansions/over all expansions. High axial shift; law vacuum /high exhaust hood temperature; rotor eccentricity high. Under such condition   operator should check gland steam temperature and maintain between 130-150 C if gland steam is from deaerator; maintain steam parameters as per recommendations; check lubricating oil pressure and temperature after oil cooler.

11.6.4. Turbine Eccentricity High

This could be due to deflection of turbine rotor either due to improper heating during start up, improper draining of steam lines and turbine cylinders contributing water entry to turbine, abrupt drop of main steam temperature due to water carry over from boiler, moisture in gland steam, deformation of cylinders, disengaged barring gear etc. High eccentricity results in high turbine vibration and unusual noise from turbine. Operator should also follow start up procedure regarding steam parameters, draining of steam lines and cylinders, soaking times etc. strictly. Operator should also maintain lubricating oil temperature between 40 °and 45°; check quality of lubricating oil, bearing temperature of running turbine etc. and keep spray control valves of bypass system properly isolated on idle turbine.

11.6.5. Water Hammer in Main Steam Line

Water hammering can result in damage to steam lines connected valves or even steam turbine. It may occur due to water carry over from boiler, inadequate draining at the time of charging main steam line. Water penetration through extraction lines; such a condition is detected by unusual noise in the system, abrupt fall in steam temperature after superheater, rapid increase in load with resulting fall is pressure in drum which is accompanied by noise in drum level; high level in deaerator, as water will enter into seals and control valves under this condition; inadequate draining of collecting condensate; and water entry from spray system of pressure reducing and desuperheatring stations; forming of water in boiler drum due to pressure of salts. Under such a condition, all drains should be wide opened; boilers tripped increase of water carry over and spray control valves on reducing stations closed and heater levels maintained.

11.6.6. Condenser Tube Failure

This is detected by high conductivity of condensate, solids, particularly chlorides getting into boiler making pH fall, and conductivity raise. This could occur due to erosion/corrosion of tubes, pitted tubes, etc. or maintain excessive cooling water inlet pressure leading to leakage from expanded joints, maintaining low cooling water flow causing high terminal difference leading to weakening of expansion joints, or slushy circulating water causing erosion. Under this condition, operator should isolate one side of condenser and make sure conductivity improves after isolation, check CW inlet pressure terminal temperature difference and regulate same as per norms, increase continuous blow down as per repairs.

11.7. Emergencies in Generator and Electrical Equipment

1. Tripping of tie in grid breaker
2. Tripping of generator circuit breakers
2. Loss of supply to unit auxiliaries
3. Loss of supply of station auxiliaries
4. Loss of excitation of the generator
5. Vibration of the generator and falling out of synchronies
6. Severe sparking under exciter commutator and generator slip ring brushes
7. Damage o generator bearing and seals
8. Hydrogen cooling system
9. Decrease in purity of hydrogen
10. Water in generator body
11. Rise in gas temperature
12. Seal oil system
13. Generator stator cooling water leakage
14. Seal oil damper tank level low
15. Stator winding temperature high
16. Over fluxing

12. POLLUTION FROM THERMAL POWER PLANTS

12.1. Introduction

India is the world's second most populous nation. Its population has grown from 300 million in 1947 to more than a billion today. Rising population and increasing rates of energy consumption, consistent with rapid economic growth and changes in lifestyles, exacerbate the process of environmental degradation in developing countries like India and China. Table 12.1 provides a comparison of the basic data for energy consumption and emissions in India, China, and the United States, as given in World development indicators. Per capita energy consumption in India in 1997 was 19.16 million BTU, compared to 323 million BTU in the United States and a world average of 64.8 million BTU. By the year 2010, per capita energy usage in India is expected to increase around 40 million BTU. India's per capita energy consumption and carbon emissions are relatively low. In 1997, India emitted 0.3 metric tons of carbon per person, approximately one quarter of the world average and 20 times less than the United States. India's low per-capita emissions are compensated by large population, making it the sixth largest greenhouse gas emitter in the world. With 4.0 percent annual growth rate in energy consumption, India is projected to be a major contributor of the GHGs and other pollutants increasingly warranting appropriate corrective measures.

Table 12.1: Basic data for India, China, and United States

Variables	India	China	United States	Units
Population	962.4	1,227	267.7	Millions
Gross domestic product	420.8	898	7,844	US$ billions
Commercial energy production	404,503	1,097,210	1,683,810	Thousand metric tons of oil equivalent[1]
Commercial energy consumption	461,032	113,050	2,162,190	Thousand metric tons of oil equivalent
Average annual percentage growth	3.9	4.0	1.4	%
Commercial energy use per capita	479	907	8076	Kg of oil equivalent
Average annual percentage growth	1.9	2.6	0.4	%
GDP per unit of energy use	4.2	3.3	3.6	PPP[2] $ per kg oil equivalent
Carbon dioxide (CO_2) emissions	1029.2	3,108.0	5,374.3	Million metric tons
Carbon dioxide emissions per capita	1.1	2.5	20.0	Metric tons

[1]*One metric ton of oil equivalent = 40 million BTU*

[2]*PPP is gross domestic product converted to international dollars using purchasing power parity rates.*

12.2. Thermal Power Plants in India

According to the Central Electricity Authority of India, as of March 31, 1998, 83 steam plants were in operation in India. These plants generated almost 80% of total generated power for the nation. Coal consumption by various plants in the country during the year 1997-98 was almost 203 million metric tons. The consumption of fuels such as furnace oil decreased by more than 32.5%, while the consumption of lignite coal, a low-sulfur heavy stock (LSHS), a high sulfur heavy stock (HHS), and diesel oil increased by 7.54%, 31.91% and 33.9%, respectively. Decreased use of furnace oil has decreased the emissions to some extent.

12.3. Emissions from Thermal Power Plants

The main emissions from coal combustion at thermal power plants are carbon dioxide (CO_2), nitrogen oxides (NO_x), sulfur oxides (SO_x), chlorofluorocarbons (CFCs), and air- borne inorganic particles such as fly ash, soot, and other trace gas species. Carbon dioxide, methane, and chlorofluorocarbons are greenhouse gases. These emissions are considered to be responsible for heating up the atmosphere, producing a harmful global environment. Oxides of nitrogen and sulfur play an important role in atmospheric chemistry and are largely responsible for atmospheric acidity. Particulates and black carbon (soot) are of concern, in addition to possible lung tissue irritation resulting from inhalation of soot particles and various organic chemicals that are known carcinogens.

CO_2, SO_2, NO, and soot emissions from each of the power plants have been computed. Emissions from combustion of the supplementary fuels such as high-speed diesel (HSD) and furnace oil used in small quantities (<1%) are not counted in the present calculations.

12.3.1. Carbon Dioxide Emissions

Utilities burn mostly coal with approximately 10–30% excess air. The total carbon obtained from analysis is converted to CO_2 after the reaction (combustion) is complete. Total CO_2 emissions for 1997 from all the power plants in India are estimated at 1.1 Teragrams (Tg) per day or 397 Tg per year. Average CO_2 emission per unit of electricity is 1.04 Gigagrams (Gg). Technological improvements in efficient combustion of coal can lead to greater production of electricity per unit of coal that will effectively reduce CO_2 emission per unit of electricity. Although the current per capita carbon dioxide (CO_2) emission in India is only one quarter of the world average and about twenty times less than the United State's averages, the growth rate of emissions is very high. Because of this growth, the region is expected to soon become a major contributor of greenhouse gases, such as CO_2 and other air pollutants.

12.3.2. Sulfur Dioxide Emissions

The sulfur content in Indian coal is low compared to United States coal. Acid rain due to sulfur dioxide emissions is presently not of great concern. However, increasing coal use or blending Indian coal with imported coal of higher calorific value (further increasing electricity production) needs to be carefully addressed through viable technological options. Average SO_2 emissions per unit of electricity are 0.0069 Gg. Total SO_2 emissions are estimated to be 7.33 Gg per day or 2.7 Tg per year.

12.3.3. Emissions of Oxides of Nitrogen

Oxidation of nitric oxide (NO) discharged in combustion products forms nitrogen dioxide (NO_x) in the atmosphere. These oxides of nitrogen are responsible for the formation of photochemical smog. Nitric oxide emission per unit of electricity is estimated at approximately 0.00056 Gg. Total No emissions are estimated to be 0.5 Gg per day and 0.185 Tg per year. Nitrogen oxides are important chemical species in the atmosphere since they contribute to its acidity; they also act as precursor gases for the formation of tropospheric ozone. Tropospheric ozone is a greenhouse gas responsible for global warming and is also known to have an adverse effect on plants. NO_x emissions should be kept at a minimum possible level. Lower concentrations of NO_x lessen the formation of tropospheric ozone even when other precursor gases like carbon monoxide (CO) are present in higher concentrations.

12.3.4. Carbonaceous Material and Black Carbon (Soot)

Incomplete and/or inefficient combustion processes of fossil fuel to generate soot. A recently conducted Indian Ocean Experiment (INDOEX) suggests that the presence of soot, carbon in the atmosphere over the northern Indian Ocean hinders its natural heating processes by about 15%. Enhancement of boundary layer heating can significantly influence regional hydrological cycles and climate. Present calculations show that soot emissions are produced at a rate of 22.0 Gg per year from Indian thermal power plants. Soot emissions in India have not been studied thoroughly so far; these are the first estimates of soot emission from Indian thermal power plants. Appropriate technological intervention to prevent soot, carbon emissions may possibly not only reduce the chances of soot escaping into the atmosphere (where it can potentially change the radiation balance), but can also lead to further increases in electricity production.

12.4. Emissions from Coal Usage

The main emissions from coal combustion at thermal power plants are Carbon dioxide (CO_2), Nitrogen oxides (NO_x), Sulfur oxides (SO_x), Chlorofluorocarbons (CFCs), carbonaceous material (soot), and airborne inorganic particles such as fly ash, also known as suspended particulate matter (SPM) and other trace gas species. Carbon dioxide, nitrous oxide, and chlorofluorocarbons are greenhouse gases. Evidence accumulated by the Inter-governmental

Panel on Climate Change (IPCC) suggests that emissions of these greenhouse gases might be responsible for climate change, a global concern. Possible consequences projected by IPCC include:

1. A rise in sea levels
2. A more vigorous hydrological cycle that may increase the severity of floods and droughts and may cause more extreme climatic events
3. Change that could threaten agricultural productivity

Oxides of nitrogen and sulfur, also play an important role in atmospheric chemistry and are largely responsible for atmospheric acidity. Particulates and black carbon (soot) are of concern in the radiative[10] forcing of the earth. They also have a significant negative impact on human health causing lung tissue irritation and are linked to cancer and other serious diseases.

The pollutants emitted from thermal power plants depend largely upon the fuel burned, the furnace design, the excess air, and any additional devices used to reduce the emissions. At present, the only control device used in thermal power plants in India is an electrostatic precipitator to control the emission of fly ash (SPM). CO_2, SO_2, nitric oxide (NO), soot, and SPM emissions from each of the thermal (coal-fired) power plants in India have been computed using basic principles of combustion. These calculations are based on a theoretical ideal and the input data, such as chemical composition of the coal used in the power plants, coal used per unit of power, excess air used during combustion, and the power generation from each plant. This input data have been collected from the published information. The present method to estimate the emissions is one of the many available methods for the emissions inventory process. The other methods used in different countries are based on the guidelines recommended by IPCC[11], and may require large resources. Emissions from combustion of the supplementary fuels such as high-speed diesel (HSD) and furnace oil used in small quantities are not counted in the present calculations.

12.4.1. Emission of Carbon Dioxide and Sulfur Dioxide:

Utilities mostly burn coal with approximately 10 -30% excess air. Carbon as obtained from Ultimate analysis is converted to CO_2 after the reaction (combustion) is complete. Some carbon is emitted in the form of soot and some carbon remains unburned and mixes with the ash. Different combustion technologies affect the types and concentrations of resultant species, e.g. fixed bed combustion results in higher carbon content in the ash.

Carbon in the coal is converted to carbon dioxide (CO_2) by the reaction

$$C \ + \ O_2 \ --> CO_2$$

Similarly, hydrogen and sulfur are converted to moisture (H_2O) and sulfur dioxide (SO_2) by the reactions

$$H_2 \quad + \quad O_2/2 \quad \text{-->} H_2O$$

$$S \quad + \quad O_2 \quad \text{-->} SO_2$$

Table 12.2, as an example, gives the computation of oxygen required for burning one kg of coal used at the Chandrapur thermal power station and the combustion products.

Table 12.2: Computation of Combustion Products for the Chandrapur Coal

Species	Mass	Oxygen required	Products	Reaction
Carbon	0.3769	0.3769 x (32/12) = 1.01	CO_2: 0.3769 x (44/12) = 1.38	$C + O_2 \text{-->} CO_2$
Hydrogen	0.0266	0.0266 x (16/2) = 0.21	H_2O: 0.0266 x (18/2) = 0.24	$H_2 + O_2/2 \text{-->} H_2O$
Sulphur	0.008	0.008 x (32/32) = 0.008	SO_2: 0.008 x (64/32) = 0.016	$S + O_2 \text{-->} SO_2$
Oxygen	0.0578			
Nitrogen	0.0107			$N_2 + O_2 \text{-->} 2NO$
Ash	0.47			

Oxygen required to burn 1 Kg of coal = 1.01+0.21+0.01−0.05 = 1.18 Kg

Air required = (Oxygen required)/(Mass fraction of oxygen in the air)

= 1.18/0.233 = 5.06 Kg

= stoichiometric air

Total air (stoichiometric + 20% excess)

= 5.06x1.2 = 6.072 Kg

6.072 air = 1.415(6.072 x 0.233) oxygen + 4.474(6.072 x 0.767) nitrogen

The combustion product with 20% excess air will contain: 0.2348(1.415−1.18) Kg O_2 and 4.48(4.47 + 0.01)Kg N_2.

Appendix-A gives the general formula for the calculation of required oxygen for burning of one kg of fuel.

Table 12.3 gives the computation of the flue gas composition from burning one Kg of Chandrapur coal at 20% excess air, as an example.

Table 12.3: Flu Gas Composition with 20% Excess Air

Product	Mass/Kg coal	Mol. wt	Kmoles/Kg coal	% volume
CO_2	1.38	44	0.03136	15.39
SO_2	0.016	64	0.00025	0.12
O_2	0.24	32	0.0075	3.68
N_2	4.61	28	0.1646	80.77
		Total =	0.2038	

12.4.2. Emissions of Oxides of Nitrogen from Coal

Oxides of nitrogen (NO_x) are (i) nitrous oxide (N_2O), (ii) nitric oxide (NO), and (iii) nitrogen dioxide (NO_2). NO_2 is mostly formed by oxidation of the NO, which is discharged in combustion products. About 90% of the NO_x is in the form of NO. NO is formed by two mechanisms: (i) oxidation of atmospheric nitrogen, known as 'thermal NO' and (ii) oxidation of nitrogen that is chemically bound within the fuel, known as 'chemical NO'. The amount of NO_x varies widely with boiler conditions. NO_x emissions are generally functions of flame temperature, excess air, percentage of boiler load, nitrogen content in the coal, and rate of gas cooling. In pulverized coal flames, about 30 - 35% of nitrogen in coal gets converted into NO and remaining nitrogen in the coal gets converted into molecular nitrogen. The actual mechanism, whereby atmospheric nitrogen is oxidized, goes through a complex chain of reactions initiated by oxygen atoms. We can however calculate equilibrium concentrations of NO, using the following reaction:

$$N_2 + O_2 \longrightarrow 2NO$$

This is a lumped reaction. Generally accepted principal reactions are

$$O + N_2 = NO + N$$

$$N + O_2 = NO + O$$

$$N + OH = NO + H$$

The concentration of nitric oxide (NO) is given by

$$X_{NO} = K_{10.1} \, (X_{N_2})^{0.5} \, (X_{O_2})^{0.5}$$

Where X is the species concentration and $K_{10.1}$ is a equilibrium constant and depends upon the temperature of the gas. Appendix-B gives the equations to compute equilibrium constant for NO reaction. Concentration values X for O_2, N_2, CO_2, SO_2, and NO for the Chandrapur coal, as calculated by the above method, are given in Table 12.4. The value of the equilibrium constant $K_{10.1}$ computed at 1700 K is 0.007824.

Table 12.4: Species Concentrations in Flue Gas for the Sample Coal

Air	$X(O_2)$	$X(N_2)$	$X(CO_2)$	$X(SO_2)$	$X(NO)$
Stoichiometric	0	0.831333	0.167148	0.001512	0
5% excess	0.010483	0.8293267	0.158754	0.001436	0.00073
10% excess	0.019970	0.8274488	0.151213	0.001368	0.001006
15% excess	0.028597	0.8257412	0.144356	0.001306	0.001202
20% excess	0.036475	0.8241817	0.138094	0.001249	0.001357
25% excess	0.043699	0.8227519	0.132352	0.001197	0.001484
30% excess	0.050345	0.821436	0.127069	0.001149	0.001591

Species concentrations (PPM) in the combustion products for the Chandrapur coal are given in Table 12.5.

Table 12.5: Species Concentrations in Parts Per Million (ppm)

Excess Air %	CO_2	SO_2	NO
0	167148	1512	0
5	158754	1436	730
10	151213	1368	1006
15	144356	1306	1202
20	138094	1249	1357
25	132352	1197	1484
30	127069	1149	1591

12.4.3. Carbonaceous Material and Black Carbon

Incomplete and/or inefficient combustion processes of fossil fuel generate carbonaceous aerosols. The emitted carbonaceous (soot) aerosols are of two types, namely organic carbon (OC) and black carbon (BC). These two types have different properties in the atmosphere. OC is a reactive species and has scattering properties in the solar spectrum[12] while the BC, on the other hand, is non reactive in the atmosphere but has highly absorbing properties in the solar spectrum . In the thermal power plants, most of the soot, carbon emitted would be in the form of BC because of the higher temperatures of combustion in the furnaces. When the soot is formed the analysis ranges from C_8H to $C_{12}H$. The importance of BC in the radiative balance of Earth is gradually being understood. The soot, carbon from the combustion of coal in the Indian thermal power plants has been calculated on the basis of prevalent combustion characteristics. It is assumed that approximately, 10% of the coal carbon goes in the bottom ash and about 2% of the carbon forms the soot that may emit with fly ash. Electrostatic Precipitators (ESPs) used in the thermal power plants in India, remove about 99% fly ash from the stack. 1% of the generated fly ash, which contains 2% soot, carbon is emitted into the atmosphere. These soot particles are sub-microns in size. Similar assumption has also been used for calculation of soot, carbon from lignite based thermal power plants in India (viz. Kutch and Neyvelli power plants).

12.4.4. Suspended Particulate Matter

The fly ash in the form of Suspended Particulate Matter (SPM) is a major pollutant from coal burning power plants in India. SPM has been calculated on the basis of the ash contents of the coal. It is assumed that 85% of the ash in the coal goes out through the stack as fly ash after the combustion.

Electrostatic precipitators (ESPs) working on 99% efficiency rate allows only 1% of the formed fly ash to emit as SPM.[10]radiative forcing is defined as a change in the average net radiation. It is based on global balance between incoming solar energy and outgoing terrestrial radiative energy that is emitted to space.

12.5. Power Plants in India

According to Central Electricity authority of India, there were 83 coal fired thermal power plants and 305 hydro plants existing in India on March 31, 1998. Thermal power (coal, diesel and gas) generation during 1997-98 was 338104 and hydropower generation was 74582 GWH. In addition to this, there were 1685 selected industrial units which had captive thermal power plants of >1MW in 1997-98. The total installed capacity of captive plants was 13004 MW and these had gross generation of 44050 GWH (i.e. 13% of the public utilities' generation) during the year 1997-98.

Table 12.6 provides basic statistics for fuel and power generation at the coal fired power plants in India.

Table 12.6: Power and Coal Usage Statistics at the Thermal Power Plants in India

Sr.No	Power Station	Installed Capacity (MW/day)	Generation (MW/day)	Coal per unit of electricity (kg/KWH)	Million KWH per day(MU)™
1R	BADARPUR	705	502	0.81	14.87
2R	I.P. STATION	277.5	136	0.86	2.96
3R	RAJGHAT	135	62	0.83	1.93
4R	FARIDABAD	165	131	0.97	2.96
5R	PANIPAT	650	362	0.82	9.74
6R	BHATINDA	440	300	0.73	6.83
7R	LEHRA	420	425	0.73	9.96
8R	ROPAR	1260	840	0.70	22.67
9R	KOTA	850	856	0.75	20.40
10R	SURATGARH#	500	516	0.73	11.85
11P	ANPARA	1630	1630	0.70	37.71
12R	HARDUAGANG	385	135	1.02	3.07
13R	DADRI	840	653	0.65	18.9
14R	OBRA	1442	1014	0.88	21.30
15R	PANKI	274	165	0.9	3.56
16R	PARICHA	220	80	0.88	1.75
17P	RIHAND	1000	960	0.66	24.09
18P	SINGRAULI	2000	1545	0.61	44.06
19R	TANDA	330	195	0.96	4.9
20R	UNCHAHAR	840	826	0.74	19.66
21	KUTCH%	215	166	0.73	3.65
22R	SABARMATI#	330	326	0.73	6.6
23	DHURVARAN	534	348	0.73	8.23
24R	GANDHI NAGAR	870	722	0.60	17.09

25R	SIKKA	240	225	0.67	4.99
26R	UKAI	850	775	0.72	16.74
27R	WANAKBORI	1470	1048	0.67	25.14
28R	BHUSAWAL	478	445	0.79	10.80
29R	CHANDRAPUR	2340	1705	0.79	44.39
30R	DAHANU	500	412	0.62	10.77
31R	KHAPER KHEDA	840	630	0.76	14.84
32R	KORADI	1080	662	0.84	17.03
33R	NASIK	910	810	0.74	17.6
34R	PARAS	58	29	0.82	0.96
35	TROMBAY	1150	658	0.73	15.61
36R	PARLI	690	625	0.80	14.73
37P	AMARKANTAK	290	201	0.72	4.7
38R	SANJAY GANDHI	840	770	0.72	17.19
39P	KORBA II&III	400	315	0.97	7.29
40P	KORBA WEST	840	630	0.68	14.48
41P	KORBA STPS	2100	2153	0.68	50.14
42R	SATPURA	1143	845	0.88	19.22
43P	VINDHYACHAL	2260	2204	0.65	51.13
44R	KOTHAGUDEM	1170	1004	1.08	22.39
45R	RAYALSEEMA	420	432	0.78	10.08
46R	NELLORE	30	23	1.21	0.45
47R	RAMAGUNDAM	63	58	0.73	1.45
48P	RGUNDAM(STPS)	2100	2160	0.63	50.65
49R	VIJAYAWADA	1260	1281	0.73	30.41
50I	ENNORE	450	157	0.94	3.75
51I	METTUR	840	860	0.75	20.33
52	NEYVELLI@	600	550		12.77
53I	NORTH MADRAS	630	646	0.76	15.2
54I	TUTICORIN	1050	1070	0.74	25.13
55R	RAICHUR	1260	850	0.76	21.2
56R	BARAUNI	310	40	1.06	1.25
57R	MUZAFFARPUR	220	65	1.09	1.5
58R	PATRATU	770	274	0.96	5.23
59R	TENUGHAT	420	200	0.87	4.8
60P	KAHALGAON	840	745	0.84	17.7
61R	BOKARO A&B	805	260	0.80	5.70
62R	CHANDRAPURA	750	100	0.79	3.39
63R	MEJIA	630	475	0.70	9.9
64R	DURGAPUR-DVC	350	300	0.62	7.2
65R	BANDEL	530	150	0.59	4.87

66R	CALCUTTA	190	141	0.89	3.07
67	New Cossipore	130	50.6		2.11
68R	S.GEN STATION	135	136	0.62	3.26
69R	DURGAPUR -DPL	390	93	0.74	4.39
70R	KOLAGHAT	1260	1210	0.73	29.0
71	BAKRESHWAR*	630	383	0.77	9.2
72R	SANTALDIH	480	150	0.60	3.6
73R	TITAGARH	240	240	0.64	5.76
74R	BUDGEBUDGE*	500	500	0.77	12
75P	FARAKKA	1600	1470	0.83	34.95
76P	TALCHER-NTPC	460	390	1.01	9.4
77P	TALCHER-STPS	1000	490	0.75	11.60
78	IB VALLEY TPS	420	406	0.91	9.70
79R	BONGAIGAON	240	0	0.66	0.8
80	Chandrapur(Assam)	60	7.82	0.66	0.33
81	Namrup	30	5.13	0.66	0.21
TOTAL	ALL INDIA	53216		0.73	

Kg/KWH are the average for the Eastern region

 # Kg/KWH are the average for the western region

@ Lignite is used as the fuel

™ Based on monthly average

R Rail feed plant

P Pit-head plant

I Inter modal

Note: Amarkantak and Amarkantak Extension are merged together.

Similarly Korba II and III are merged together.

Coal consumption by various plants in the country during the year 1997-98 was 202.75 million metric tones. The specific coal consumption in 1997-98 is 0.73 Kg/KWH. Average heat input for all the power plants was 2801 Kcal/KWH in 1997-98. Some power plants are more efficient than others. The efficiency depends on the combustion technology, operating conditions, and the coal properties. Out of 81 coal fired thermal power plants, 13 power plants, having aggregate installed capacity of 7594 MW, produced 43387 GWH. This represents 11% of total steam generation and an overall thermal efficiency of less than 25%. 42 power plants with an aggregate installed capacity of 27080 MW had an overall thermal efficiency between 25% to 30% and generated 183558 GWH, which is 46% of total steam generation. 26 power plants with aggregate capacity of 24850 MW had an overall thermal efficiency above 30% and produced 171349 GWH representing 43% of total steam generation.

12.6. Emissions from Coal Fired Thermal Power Plants in India

Based on the input parameters and the ultimate analysis of coal used for power generation, emissions of CO_2, SO_2, NO, soot carbon and particulate matter from each of the power plants has been computed. Input parameters (operating conditions) are actual air supplied, electric power generated per day, and coal used for unit power generation. Thermal power plants also use small quantities of diesel oil and furnace oil (FO) as supplementary fuels to boost the combustion and heat content. In the thermal power plants run by National Thermal Power Corporation (NTPC), supplementary fuel consumption is 0.2 to 0.3 ml/unit of power. The supplementary fuel consumption in old thermal power plants may range from 1 to 4% of the fuel. Emissions from combustion of these supplementary fuels are not accounted in the computations at present used for unit (kwh) power generation, For the estimation of emissions of above mentioned species from the Indian thermal power plants, the available values of the ultimate analysis of coals used in the seven thermal power plants namely Chandrapur, Dhanau, Singrauli, Dadri, Rihand, Kutch, and Neyveli, are used. Most of thermal power plants in India use E and F grade coals only. The excess air used in the individual power plants, kg coal and per day power generation is available and is listed. NO calculations assume equilibrium reactions and 1700 K gas temperature.

CO_2 emissions are estimated based on the carbon content in the coal and the excess air used at the power plants. 12% carbon (based on the measurement data, it is assumed that 10% carbon remains unburned and mixes with the ash and 2% carbon forms soot) is subtracted before calculating emissions of CO_2. Table 12.7 provides CO_2 emissions per day from power plants.

Table 12.7: Computed Values of CO_2 Emission Per Day from Power Plants

CO2 (Thousand ton)	Number of power plants	Coal used per day (million tons)
40-50	5	0.16
30-40	3	0.08
20-30	9	0.15
10-20	24	0.3
0-10	40	0.43

Total CO_2 emissions per day from all the coal fired power plants in India was 1.1 thousand tons per day in 1997-98 and annual emission has been computed to be 395 million tons. Yearly estimates of power generation and emissions are based on the plant load factor (PLF) for individual power plant. PLF is the ratio of the actual power generation and the installed capacity. Knowing the yearly power generation, the annual emissions are computed. Estimate of CO_2 emission from power sector in India for 1990 is 213 million tons. CO_2 emission estimates based on the present calculations seem to compare well with this estimate made in 1990. CO_2 emissions per unit of electricity from power plants in tabular form are given in Table 12.8.

Table12.8: Computed Values of CO_2 Emission Per Unit of Electricity from Power Plants

CO_2 (Kg/KWH)	Number of power plants
1.4 - 1.8	4
1.2 - 1.4	9
1.0 - 1.2	27
0.8 - 1.0	40
0.6 - 0.8	1

Computed data indicate that CO_2 emissions per unit of electricity from most power plants range between 0.8 and 1.2 kg/KWH. Some plants have between 1.2 and 1.4 kg/KWH. Four plants have CO_2 emissions more than 1.4 kg/KWH. This number reflects operational inefficiency due to poor coal quality or operating conditions. One power plant has this number less than 0.8 kg/KWH.

For the soot, present model calculations based on the assumptions given before, show that approximately 67 tons soot, carbon (BC) per day or about 24 thousand tons per year is emitted from the Indian thermal power plants during 1997-98. These estimates are much lower than 608.4 thousand tons of BC emissions estimated from coal consumption in India based on average emission factor[15]. This has the implied consideration that the thermal power plants are the major consumers of coal in India. The emission factors arising out of the present calculations are 0.08 gm/kg of coal and 0.06 gm/kg of lignite in Indian thermal Power. This emission factor for coal is much lower than average emission factors for BC of 1.0, 0.325 and 0.2 g/kg proposed for under-developed, semi-developed and developed countries respectively[14] for industrial use of hard coal, hard coal briquettes, coke, oven coke, gas coke and brown coal coke. However, Soot emission factor of 0.08 gm/kg for coal obtained from the present calculations compares well with the emission factor of 0.075 proposed for industrial use of hard coal[16,17]. The Nellore thermal power plant with an estimated emission of 0.1 gm/KWH has been found to be the largest emitter of soot. The other large emitter thermal power plants include Faribdabad, Harduaganj, Korba II& III, Kothagudem, Barauni, Muzaffarpur and Talchar NTPC where the soot emission range from 0.08 to 0.1 gm/KWH.

The ash contents in the coal consumed in the thermal power stations is responsible for the emission of 2.3 thousand tons of SPM in the form of fly ash per day and about 0.8 million tons SPM per year. The Chandrapur, Kothagudem, Nellore, Baauni and Muzaffarpur thermal plants have been found to be amongst the largest emitter of SPM per unit electricity. About 3 to 3.5 gm of SPM/KWH has been estimated to have emitted from these plants. The plants like Faridabad, Harduaganj, Obra, Panki, Paricha, Tanda, Korba II & III, Satpura, Ennore, Patratu, Calcutta, New Cossipore, Talchar NTPC and IB Valley TPS emit in the range of 2.5 to 3 gm SPM/KWH. Rest of the plants have SPM emissions lower than 2.5 gm/KWH.

Appendix A

Oxygen required (OR) for burning one Kg of fuel

$$= C \times (32/12) + H \times (16/2) + S \times (32/32) - O_2$$

Stoichiometric air for burning one Kg of fuel = OR/0.233

Where C, H, S, and O_2 are species mass per kg of fuel.

Appendix B

Calculation of equilibrium constant for the reaction

$N_2 + O_2 \rightarrow 2\ NO$

$DG^0 = -RT \ln K_0$

Where DG^0 is the standard free energy of formation. K_0 is the equilibrium constant at temperature T.

The most reliable and easily accessible values are those tabulated for T = 25°C = 298 K. DG^0 = 20.69 Kcal/mol for NO formation at this temperature.

$K_0 = \exp(-DG^0 / RT)$

van' t Hoff equation:

$d(\ln K)/dT = DH^0 / RT^2$

where $DH^0 - DG^0 = DS^0$

$\ln(K_1 / K_0) = DH^0 / R \times (1/T_0 - 1/T_1)$

$K_1 = K_0 \times \exp(DH^0 / R \times [1/T_0 - 1/T_1]$

DH^0 = 21.57 Kcal/mol for NO formation. This gives $K_0 = \exp(-34.89)$ and K_1 = 0.000526 for 1200 K.

12.7. Energy Activities

Combustion of fossil fuels is the main source of energy and is largely responsible for polluting the natural environment. India is the world's seventh largest energy consumer. Coal is the primary fuel for thermal power plants. Gasoline and diesel are the primary fuels for vehicular transport. There is also a limited usage of natural gas. USDOE has estimated total primary energy consumption and carbon dioxide emissions from fossil fuels in India as given in Table 12.9. The Government of India is under the process of preparation of its Initial Communication, which would provide the data for CO_2 emission from coal usage.

Table 12.9: Primary Energy Consumption and CO_2 Emissions

Fossil fuel	1996	1997	1998	Units
Total Primary Energy Consumption	11.60	12.10	12.51	10^{15} BTU
Coal Consumption	6.49	6.70	6.88	10^{15} BTU
Petroleum Consumption	1681	1765	1840	Thousand Barrels per day
Natural Gas Consumption	0.80	0.83	0.88	10^{15} BTU
CO_2 emissions from coal	591.87	611.01	627.48	Million metric tons carbon
CO_2 emissions from petroleum	223.81	237.42	247.6	Million metric tons carbon
CO_2 emissions from natural gas	47.19	48.47	51.11	Million metric tons carbon

12.8. Emissions from Coal Fired Thermal Power Plants

Coal is the primary fuel for energy in India, followed by petroleum, and natural gas. India is third-largest producer of coal in the world. According to Geological Survey of India's estimates, coal reserves were 212 billion tones as on January 1,2000. The state-wise distribution of coal reserves and coal types (as per their geological formation) are given in table 12.10.

Table 12.10: Coal Reserves (Million tones) in India

State	Proved[3]	Indicated[4]	Inferred[5]	Total	Coal Type
Andhra Pradesh	7346	3312	2,929	13,587	Gondwana
Arunachal	31	11	48	90	Tertiary
Assam	259	27	34	320	Tertiary
Jharkhand	34,794	28,692	5,642	69,128	Gondwana
Jammu & Kashmir	-	-	-	90,000	Lignite
Chhatisgarh	13,010	22,148	8,334	43,430	Gondwana
Maharashtra	4,149	1,323	1,605	7,077	Gondwana
Meghalaya	118	41	301	459	Tertiary
Nagaland	4.0	1	15	20	Tertiary
Orissa	11,140	22,755	16,554	50,449	Tertiary
Uttar Pradesh	766	296	0	1,062	Gondwana
West Bengal	10,779	10,894	4,236	25,909	Gondwana
Tamil Nadu	-	-	-	24,575	Lignite
Gujrat	-	-	-	465,000	Lignite
Rajsthan	-	-	-	1,020,000	Lignite
Total	82,396	89,501	39,697	-	-

Central pollution control board: Environmental Standards

Guidelines for Discharge Point

1. The discharge point shall preferably be located at the bottom of the water body at midstream for proper dispersion of thermal discharge.
2. In case of discharge of cooling water into the sea, proper marine outfall shall be designed to achieve the prescribed standards. The point of discharge may be selected in consultation with concerned State Authorities/NIO.
3. No cooling water discharge shall be permitted in estuaries or near ecologically sensitive areas such as mangroves, coral reefs/spanning and breeding grounds of aquatic flora and fauna

13. THERMAL POWER PLANT ECONOMICS

13.1. Power Plant

The power plant is an assembly of equipment that produces and delivers mechanical and electrical energy. Electrical equipments and a power station includes generators, transformers, switch gears and control gear, Fig. 13.1 shows the main part of a power system.

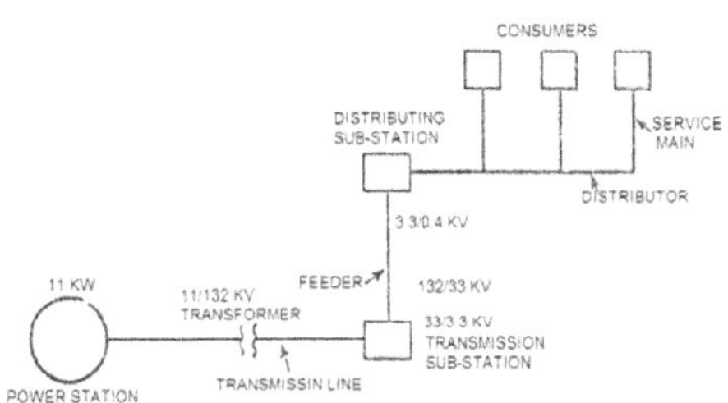

Fig. 13.1: Main Part of Power System

From the economic point of view, it is desirable that when large amounts of electric power is to be transmitted over long distance it should be transmitted at a voltage for transmission should be so chosen that it gives better efficiency, regulation and economy. Step up transformer is used to step up the generation voltage to transmission voltage, which is usually 132 KV. At the transmission substation the voltage is stepped down to medium voltage, usually 33 or 3.3KV. The feeders carry the power to the distribution sub-station. Feeders should not be tapped for direct power supply. The function of transformers at the distribution sub-station is to step down the voltage to a low distribution voltage which is usually 400 to 230 V. Distribution are used to supply power to the consumers. Transmission of electric power over long distances can be done most economically by using extract high voltage (E.H.V) lines. In the world today, many A.C extra high voltage lines are in operation. These E.H.V lines operate at a voltage higher than the high voltage lines i.e. 230kV. The E.H.V lines are now in operation in Europe, USA and Canada and operate at 330kV, 400kV, 500kV and 700kV. Still higher voltage 1000kV levels are in the experimental stage. In India there are no E.H.V lines so far, but it is hoped that soon such lines in the form of Super Grid will be developed.

13.2. Requirement of Thermal Power Plant Design

The factors to be kept in view while designing a power station are follows. Economy of expenditure i.e. minimum

1. Capital Cost

2.   Operating and maintenance cost

3.   Safety of plant and personnel

4.   Reliability

5.   Efficiency

6.   Ease of Maintenance

7.   Good working conditions.

8.   Minimum transmission loss

13.3. Useful Life of a Power Plant

Every power plant wears as the time proceeds and it becomes less fit for use. The deterioration of the equipment takes place because of age of service, wear and tear and corrosion. By a thorough program of preventive maintenance and repairs, it is possible to keep the power station in good conditions to get proper service from it. A power plant becomes obsolete when it can be replaced by one of more modern design which operates at a reduction in total annual costs. Therefore, useful life of a power plant is that after which repairs becomes so frequent and expensive that it is found economical to replace the power plant with a new one. The useful life of a conventional thermal power plant is 20 to 25 years, the useful life of nuclear power plant is 15 to 20 years and the useful life of a diesel power plant is about 15 years. The useful life of some of the equipment of a steam power plant is indicated in table 13.1.

Table 13.1: The Useful Life of Some of the Equipment of a Steam Power Plant is Indicated in Table

Equipment	Useful life (Years)
Steam turbine	5-20
Steam turbo-generators	10-20
Boiler 1.fire tube 2.water tube	10-15 20
Coal and ash machinery	10-20
Pumps	15-20
Feed water heaters	20-30
Condensers	20
Stacks	10-30
Strokes	10-20
Transmission lines	10-20
Transformers	15-20
Motors	20
Electric meters and instruments	10-15

13.4. Load Forecasting

Preparation of forecasts of future electric demands is a continuous process and is, necessarily, a matter of judgment applied to the interpretation of the collected records of the previous demand values, economic trends, population shifts, technology changes and related matters. In a developed country like ours socio–economic consideration and the imminent need for equitable distribution of benefits of available national resources to all regions will also constitute an essential factor. The effectiveness of planning of the growth of power industry as also its efficient operation is largely dependent on the accuracy with which forecasts of load are made. The growth rate of demand is also dependent upon the actual growth rate of generating capacity, its availability and its dependability. In many regions of the country the growth had been throttled for want of the power industry keeping to its promised growth rate.

There are two significant characteristics of electrical demand (usually called "load") viz. the peak load which is the maximum demand on the generating plant for a relatively short period of 15 minutes or one hour; and the average demand over a longer period, usually a month or a year. As already indicated, the ration between the average demand and the peak load is called the load factor. The higher the load factor, the more profitably the installed capacity is utilized and such a system can afford to invest more money initially for more efficient equipment that will reduce operating costs. Some of the factors considered in forecasting the load growth are:

1. The overall trend of the rate of increase in the demand for energy/power over a period.
2. The trend of the rates of growth of various components of load which when synthesized yields the total system requirements.
3. The pattern of seasonal variation in the peak demand during the year.

13.5. Size of Unit

The cost of electricity can be kept low by installation of large sized units. The advantages accruing out of large sized units are:

1. Improvement of thermal efficiency of turbine.
2. Reduction in capital cost of installation.
3. Reduction in operation and maintenance costs.

The improvement of thermal efficiency of turbine unit is possible by using higher steam pressure and higher stem temperature at the turbine inlet. Fig. 13.2 Shows the improvement in heat rate that is obtained by increasing the turbine inlet stem pressure.

The highest steam pressure used in the country remains at 130/160 kg/cm² and in the case of 500 MW units which is being imported a parameter of 170 kg/cm² is being envisaged. The temperature of steam remains at 535ºC.These is sub critical pressure. Higher pressure and temperature would involve improved materials.

13.6. Reduction in Operation and Maintenance Costs

The running expenses which constitute interest, operation and maintenance and depreciation charges are proportional to the cost per KW installed and so these changes would vary the same as the cost. per KW installed capacity. The fuel costs also would be reduced on account of the improved heat rate of large sized units.

The above advantages accruing out of increased size in the unit rating would be required to be balanced against the following:

1.   Increased reserve capacity required with large units.
2.   Increased magnitude of forced outages.
3.   Limitation in the site availability.

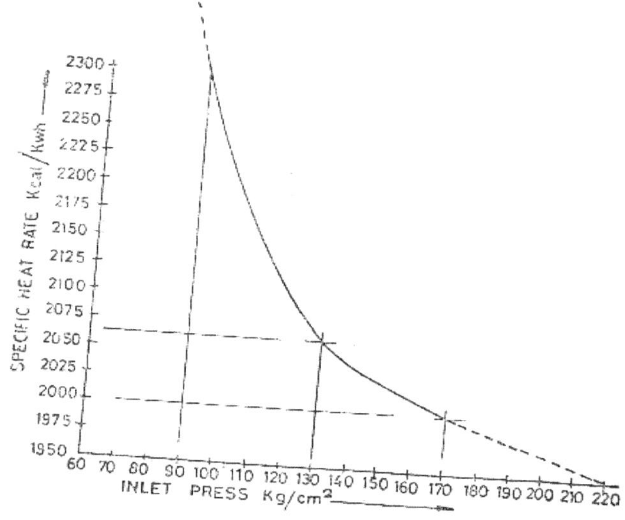

Fig. 13.2: Specific Heat Rate V$_s$ Inlet Pressure

4.   Possible increase in transmission system costs.
5.   Transport restrictions, if any.

13.7. Cost of Generation

The main factors making up the cost of generation, i.e., the cost per Kwhr at the generating stating bus bars in a thermal power station can be broadly classified as:

1. Fuel cost
2. Fixed cost
3. Operation and maintenance charges

The fuel cost covers the cost, including duties required to be paid to the coal mines, cost of transportation, cost of supporting fuel oil, coal handling costs within then power station etc.

13.8. Thermal Efficiency

The input to the power generation cycle is the heat value of fuel and the output is the electrical energy. The ratio between the gross energy output and the total energy input is called the thermal efficiency of the plant. It is not worthy that the thermal efficiency of the conventional thermal generation units is or the order of 23 to 25 percent. The most efficient station in U.K. yielded 35.36 percent during 1976-77. The station has 4x500 MW units using steam parameters 158.6 bars and 566ºC and operated at 79.4 percent load factor.

13.9. Plant Availability

Plant availability is expressed in terms of total number of hours that a unit was actually in operation with its circuit breaker closed to the station bus bars out of the total hours in a year.

13.10. Capacity Factor

Capacity factor is the ration of the number of units (KWh) actually generated by the unit in a year to the number of units that would have been generated had the unit worked at rated capacity throughout. The capacity factor has been steadily improving in the country's thermal station.

13.11. Utilization Factor

The utilization factor is expressed in terms of KWh per KW per year for an installation or for a system. In the country this factor has so far been on an average 3500KWh/KW. It is evident that if the units are operating in an interconnected system facilitation integrated operation, the utilization factor could improve all around.

13.12. Requirement of thermal power station

The major inputs to a thermal power station are coal, water, land, and transmission outlet and infrastructure covering approach road/rail, skilled labour, and suitable construction material, etc. Between coal and water it is debatable as to which of the two deserves more consideration. It could be said coal could be relegated to second place. At the same time, from the process point of view, the fuel should come first. The following are required for the thermal power plant

1. Coal
2. Water
3. Ash disposal
4. Land
5. Transmission network
6. Infrastructure

13.13. Cost of Electrical Energy

The cost of the electrical energy generated consists of fixed cost and running cost.

13.13.1. Fixed Cost

The fixed cost is the capital invested in the installation of complete plant. This includes

1. Land and building equipment
2. Interest
3. Depreciation cost
4. Insurance
5. Management cost

13.13.2. Operating Cost

The operating cost of the electrical power generation includes

1. Fuel cost
2. Oil, grease
3. Cooling water
4. Consumable articles

13.14. Selection of Type of Generation

Power supply comes from a utility, which is financed by central government, the objectives of electric–generation investment must be optimum or more efficient use of national resources. Once the total energy needed per year demand is known, the problem of selecting the type of generation arises. Theoretically, this could be achieved by steam power plant. The elimination of some of these alternatives is done on the basis of other controlling factors such as non-availability of water power sites. The points, which are to be considered in choosing the type of generation, are listed below:

1. The type of fuel available or availability of suitable sites for water power generation.

2. The cost of transmitting the energy

3. The cost of fuel transportation.

4. The cost of the foundation and land required.

5. The availability of an adequate supply of cooling water.

6. The type of load to be taken by the power plant.

7. Reliability in operation.

8. The life of the plant.

13.15. Tariffs for Electrical Energy

The rates of energy sold to the consumers depend on the type of consumers as domestic, commercial and industrial. The rates depend upon the total energy consumed and the load factor of the consumer. Whatever may be the type of consumers, all forms of energy rates must cover the following items:

1. Recovery of capital cost invested in the generating power plant.

2. Recovery of the running costs as operating costs, maintenance cost, metering the equipment cost, billing cost and many others.

3. Satisfactory profit on the invested capital as the power plant is considered a profitable business for the government.

QUESTIONS

1. A closed vessel made of steel and used for the generation of steam is called a

   (a) steam boiler   (b) steam turbine

   (c) steam condenser   (d) steam injector

2. A good steam boiler is one which produces maximum quantity of steam with the given fuel.

   Agree ( )   Disagree ( )

3. The selection of type and size of a steam boiler depends upon

   (a) the power required and working pressure

   (b) the geographical position of the power house

   (c) the fuel and water available

   (d) all of the above

   (e) none of the above

4. In a fire tube boiler, the water is contained inside the tubes which are surrounded by flames and hot gases from outside.

   True ( )   False ( )

5. Which of the following is a fire tube boiler ?

   (a) Lansashire boiler   (b) Babcock and Wilcox boiler

   (c) Yarrow boiler   (d) none of the above

6. Fire tube boilers are limited to a maximum working pressure of

   (a) 1.7 kg/cm^2 (b) 17 kg/cm^2

   (c) 170 kg/cm^2 (d) 200 kg/cm^2

7. Which of the following is a water tube boiler?

   (a) Lancashire boiler   (b) Babcock and Wilcox boiler

   (c) Locomotive boiler   (d) Cochran boiler

8. In fire tube boilers

   (a) Water passes through the tubes which are surrounded by flames and hot gases

   (b) The flames and hot gases pass through the tubes which are surrounded by water

   (c) Forced circulation takes place

   (d) None of the above

9. In water tube boilers, the water passes through the tubes which are surrounded by flames and hot gases.

   Yes ( )   No ( )

10. Fire tube boilers are

    (a) internally fired   (b) externally fired

    (c) internally as well as externally fired

11. Water tube boilers are
    (a) internally fired   (b) externally fired
    (c) internally as well as externally fired
12. Cornish boilers are multi-tubular boilers.
    True ( )   False ( )
13. In natural circulation steam boilers, the circulation of water is by convection currents which are
    set up during the heating of water.
    Right ( )   Wrong ( )
14. In forced circulation steam boilers, the force is applied
    (a) to draw water   (b) to circulate water
    (c) to drain off the water   (d) all of. the above
15. In a La-mont boiler, there is a _______ " circulation of water and steam.
    (a) forced   (b) natural
16. Lancashire boiler is a stationary fire tube boiler.
    Agree ( )   Disagree ( )
17. Water tube boilers produce steam at a_______   ". pressure than that of fire tube boilers.
    (a) lower   (b) higher
18. The rate of flow of steam is________ in case of fire tube boilers.
    (a) less   (b) more
19. Which of the following statement is correct?
    (a) A fire tube boiler occupies less space than a water tube boiler, for a given power.
    (b)Steam at a high pressure and in large quantities can be produced with a simple
    vertical boiler.
    (c) A simple vertical boiler has one fire tube.
    (d) all of the above
    (e) none of the above
20. The water tubes in a simple vertical boiler are
    (a) horizontal   (b) vertical
    (c) inclined
21. A Cochran boiler is a horizontal multi-tubular boiler.
    Yes ( )   No( )
22. The fire tubes in a Cochran boiler usually have diameter.
    (a) 6.25 mm   (b) 62.5 mm   (c) 72.5 mm
23. A grate, in a boiler, is a place in the combustion chamber upon which fuel (wood or coal) is
    burnt.
    True ( )   False ( )

24. The shell of the cochran boiler is made hemispherical

(a) to give maximum space and strength

(b) to withstand the pressure of steam inside the boiler

(c) both (a) and (b)

(d) none of the above

25. The crown of the fire box is made hemispherical in order to resist intense heat in the fire box.

True ( )   False( )

26. The fire tubes in a cochran boiler are

(a) horizontal   (b) vertical   (c) inclined

27. A single ended Scotch marine steam boiler may have one to four furnaces which enter from front end of the boiler.

Right ( )   Wrong( )

28. The fire tubes in a Scotch marine boiler are

(a) horizontal   (b) vertical   (c) inclined

29. Lancashire boiler is

(a) stationary fire tube boiler

(b) horizontal boiler

(c) internally fired boiler

(d) natural circulation boiler

(e) all of the above

(f) none of the above

30. For the same diameter and thickness of tube, a water tube boiler has------------heating surface as compared to fire tube boiler.

(a) more   (b) less   (c) equal

31. Lancashire boiler is used where working pressure and power required are

(a) low   (b) very low   (c) moderate

(d) high   (e) very high   (f) none of the above

32. The cylindrical shell of a Lancashire boiler has diameter from

(a)1 to 2 m   (b)1.25 to 2.25 m

(c)1.5 to 2.5 m   (d)1.75 to 2.75 m

33. The length of Lancashire boiler varies from

(a) 5 to 6 m   (b) 6 to 7 m

(c) 7.25 to 9 m   (d) 9 to 10 m

34. Lancashire boiler has______internal flue tubes.

    (a) one   (b) two

    (c) three   (d) four

35. The diameter of internal flue tubes of a Lancashire boiler is about -------- that of its shell.

    (a) one,-fourth   (b) two-fifth

    (c) one-third   (d) one-half

36. The internal flue tubes of a Lancashire boiler are reduced in diameter at the back to provide access to the lower part of the boiler.

    True( )   False( )

37. The damper in a steam boiler is provided to control the draught (i.e. rate of flow of air) and thus regulate the rate of generation of steam.

    Correct( )   Incorrect( )

38. Which of the following statement indicates the difference between Cornish boiler and Lancashire boiler?

    (a) Cornish boiler is a water tube boiler whereas Lancashire boiler is a fire tube boiler.

    (b) Cornish boiler is a fire tube boiler whereas Lancashire boiler is a water tube boiler.

    (c) Cornish boiler has one flue tube whereas Lancashire boiler has two flue tubes.

    (d) Cornish boiler has two flue tubes whereas Lancashire boiler has one flue tube.

    (e) none of the above

39. The diameter of Cornish boiler varies from

    (a) 0.5 to 1 m   (b) 1 to 2 m   (c) 1.25 to 2.5 m   (d) 2 to 3 m

40. The length of Cornish boiler varies from

    (a) 2 to 4'5 m   (b) 3 to 5 m

    (c) 5 to 7'5 m(d) 7 to 9 m

41. The diameter of flue tube in a Cornish boiler is--------- that of the shell.

    (a) one-fourth   (b) two-fifth

    (c) one-third   (d) three-fifth

42. Locomotive boiler is a

    (a) single tube, horizontal, internally fired and stationary boiler

    (b) single tube, vertical, externally fired and stationary boiler

    (c) multitubular, horizontal, internally fired and mobile boiler

    (d) multitubular; horizontal, externally fired and stationary boiler

43. Locomotive boiler produces steam at a very high rate.

Yes ( )   No ( )

44. The shell diameter of a Locomotive boiler is

(a) 1 m   (b) 1.5 m

(c) 2 m   (d) 2.5 m

45. The length of shell of a Locomotive boiler is

(a) 1m   (b) 2 m

(c) 3 m   (d) 4 m

46. The Locomotive boiler has

(a) 137 fire tubes and 44 superheated tubes

(b) 147 fire tubes and 34 superheated tubes

(c) 157 fire tubes and 24 superheated tubes

(d) 167 fire tubes and 14 superheated tubes

47. The maximum steam pressure in a Locomotive boiler is limited to

(a) 1.8 kg/cm^2 (b) 18 kg/cm^2

(c) 180 kg/cm^2 (d) 280 kg/cm^2

48. The fire tubes in a Locomotive boiler has -------- diameter

(a) 4.75 mm (b) 5.47 mm

(c) 7.45 mm   (d) 47.5 mm

49. The diameter of superheated tubes in a locomotive boiler is

(a) 1.3 cm   (b) 3.1 cm

(c) 13 cm   (d) 23 cm

50. Locomotive boiler is a _____boiler.

(a) fire tube   (b) water tube

51. Babcock and Wilcox boiler is a stationary type of water tube boiler.

Correct ( )   Incorrect( )

52. The water tubes in a Babcock and Wilcox boiler are

(a) vertical   (b) horizontal   (c) inclined

53. La-mont boiler, is a high pressure water tube steam boiler working on forced circulation.

Agree( )   Disagree( )

54. Which of the following boiler is best suited to meet the fluctuating demand of steam?

　(a) Locomotive boiler　　　　　(b) Lancashire boiler

　(c) Cornish boiler　　　　　　(d) Babcock and Wilcox boiler

55. Which of the following statement is correct?

　(a) Lancashire boiler is a fire tube boiler.

　(b) Fire tube boilers are internally fired.

　(c) Babcock and Wilcox boiler is a water tube boiler.

　(d) all of the above

　(e) none of the above

56. Which of the following statement is wrong?

　(a) Locomotive boiler is a water tube boiler.

　(b) Water tube boilers are internally fired.

　(c) La-moot boiler is a low pressure water tube boiler.

　(d) all of the above

　(e) none of the above

57. The fitting mounted on the boiler for its proper and safe functioning is a

　(a) water level indicator　　　(b) Pressure gauge

　(c) safety valve　　　　　　　(d) blow off cock

　(e) feed check valve　　　　　(f) fusible plug

　(g) all of the above　　　　　(h) none of the above

58. The device attached to the steam chest for preventing explosions due to excessive internal pressure of steam is called

　(a) safety valve　　　　　　　(b) water level indicator

　(c) pressure gauge　　　　　　(d) fusible plug

59. The water level indicators in a boiler are generally______in number.

　(a) one　　　　　　(b) two　　　　　(c) three　　　　　(d) four

60. The function of a safety valve is

　(a) to blow off steam when the pressure of steam inside the boiler exceeds the working pressure

　(b) to indicate the water level inside the boiler to an observer

　(c) to measure pressure of steam inside the steam boiler

　(d) to measure pressure of steam inside the steam boiler

　(e) none of the above

61. A safety valve usually employed with stationary boilers is

 (a) lever safety valve

 (b) dead weight safety valve

 (c) high steam and low water safety valve

 (d) all of the above

 (e) none of the above

62. A safety valve mainly used with locomotive and marine boiler is

 (a) lever safety valve

 (b) dead weight safety valve

 (c) high steam and low water safety valve

 (d) spring loaded safety valve

 (e) all of the above

 (f) none of the above

63. The principal function of a stop valve is to

 (a) control the flow of steam from the boiler to the main pipe and to shut off the steam completely when required

 (b) empty the boiler when required and to discharge the mud, scale or sediments which are accumulated at the bottom of the boiler

 (c) put off fire in the furnace of the boiler when the level of water in the boiler falls to an unsafe limit

 (d) increase the temperature of saturated steam without raising its pressure

 (e) heat feed water by utilising the heat in the exhaust flue gases before leaving through the chimney

64. Blow off cock in a boiler is used to

 (a) control the flow of steam from the boiler to the main pipe and to shut off the steam completely when required

 (b) empty the boiler when required and to discharge the mud, scale nr sediments which are accumulated at the bottom of the boiler

 (c) put off fire in the furnace of the boiler when the level of water in the boiler falls to an unsafe limit

 (d) increase the temperature of saturated steam without raising its pressure

 (e) heat feed water by utilising the heat in the exhaust flue gases before leaving through the chimney

65. A ____ in a boiler is used to put off in the furnace of the boiler when the level of water in the boiler falls to an unsafe limit.

(a) blow off cock   (b) stop valve

(c) fusible plug   (d) safety valve

66. The function of a superheater is to heat the feed water by utilising the heat in the exhaust flue gases before leaving through the chimney.

Yes ( )   No( )

67. An economiser is used to increase the temperature of saturated steam without raising its pressure.

Right( )   Wrong( )

68. A device used in a boiler to control the flow of steam from the boiler to the main pipe and to shut off the steam completely when required, is known as

(a) blow off cock   (b) superheater

(c) economizer   (d) fusible plug

(e) stop valve   (f) none of the above

69. A device used to empty the boiler whenever required and to discharge the mud, scale or sediments which are accumulated at the bottom of the boiler, is called blow off cock.

Yes ( )   No ( )

70. A device used to put off fire in the furnace of the boiler when the level of water in the boiler falls to an unsafe limit, is called

(a) blow off cock   (b) stop valve

(c) superheater   (d) economizer

(e) none of the above

71. A device used to increase the temperature of saturated steam without raising its pressure, is called

(a) blow off cock   (b) fusible plug

(c) superheater   (d) stop valve

(e) economiser   (f) none of the above

72. A device used to heat feed water by utilising the heat in the exhaust flue gases before leaving through the chimny, is known as

(a) blow off cock   (b) fusible plug

(c) superheater   (d) stop valve

(e) economiser   (f) none of the above

73. Which of the following are boiler accessories ?

(a) safety valve   (b) pressure gauge

(c) economiser   (d) superheater

(e) both (a) and (b)   (f) both (c) and (d)

74. An economizer _______ the steam raising capacity of a boiler.

(a) increases (b) decreases   (c) has no effect on

75. The pressure of feed water has to be raised before its entry into the boiler. The pressure is raised by a device known as

(a) feed pump   (b) feed check valve

(c) injector   (d) pressure gauge

76. There is about 15 to 20% of coal saving when an economiser is used in a boiler.

True( )   False( )

77. The power of a boiler may be defined as

(a) the ratio of heat actually used in producing the steam to the heat liberated in the furnace

(b) the amount of water evaporated or steam produced in kg per kg of fuel burnt

(c) the amount of water evaporated from and at $100°$ C into dry and saturated steam

(d) the evaporation of 15.653 kg of water per hour from and at $100°$ C

(e) none of the above

78. The evaporative capacity of a boiler is defined as the evaporation of 15'653 kg of water per hour from and at $100°$ C.

Yes ( )   No ( )

79. The amount of water evaporated or steam produced in kg per kg of fuel burnt is termed as evaporative capacity of a boiler.

Agree ( )   Disagree ( )

80. The equivalent evaporation is defined as

(a) the ratio of heat actually used in producing the steam to the heat liberated in the furnace

(b) the amount of water evaporated or steam produced in kg per kg of fuel burnt

(c) the amount of water evaporated from and at $100°$ C into dry and saturated steam

(d) the evaporation of 15.653 kg of water per hour from and at $100°$ C

(e) none of the above

81. The ratio of heat actually used in producing the steam to the heat liberated in the furnace, is known as

(a) Equivalent evaporation   (b) factor of evaporation

(c) Boiler efficiency   (d) one boiler horsepower

82. One boiler horsepower is defined as

    (a) the ratio of heat actually used in producing the steam to the heat liberated in the furnace

    (b) the amount of water evaporated or steam produced in kg per kg of fuel burnt

    (c) the amount of water evaporated from and at 100° C into dry and saturated steam

    (d) the evaporation of 15.653 kg of water per hour from and at 100° C

    (e) none of the above

83. The amount of water evaporated or steam produced in kg per kg of fuel burnt is known as
    power of a boiler.

    True ( )   False ( )

84. The ratio of heat actually used in producing the steam to the heat liberated in the furnace is
    known as equivalent evaporation from and at 100° C.

    Correct ( )   Incorrect ( )

85. The evaporation of ______of water per hour from and at 100ºC is called one boiler horsepower.

    (a) 13.5 kg   (b) 15.653 kg

    (c) 16.65 kg   (d) 17.356 kg

86. The amount of water evaporated from and at ______ into dry and saturated steam is called
    equivalent evaporation from and at 100° C.

    (a) 0° C   (b) 40° C

    (c) 60° C   (d) 100° C

87. The amount of water evaporated in kg per kg of fuel burnt is called

    (a) Equivalent evaporation from and at. 100° C

    (b) evaporative capacity of a boiler

    (c) boiler efficiency

    (d) one boiler horsepower

    (e) none of the above

88. The amount of water evaporated from and at 100° C into dry and saturated steam is called
    evaporative capacity of a boiler.

    Yes ( )   No ( )

89. The evaporation of 15.653 kg of water per hour from and at 100° C is called

    (a) equivalent evaporation from and at 100° C

    (b) evaporative capacity of a boiler

    (c) boiler efficiency

    (d) one boiler horsepower

    (e) none of the above

90. Which of the following statement is wrong?

   (a) The evaporation of 15.65a kg of water per hour from and at 100° C is called one boiler horsepower.

   (b) The amount of water evaporated in kg per kg of fuel burnt is called equivalent evaporation from and at 100° C.

   (c) The ratio of heat actually used in producing the steam to the heat liberated in the furnace is called boiler efficiency.

   (d) all of the above

   (e) none of the above

91. The main object of a boiler trial is

   (a) to determine the generating capacity of the boiler

   (b) to determine the thermal efficiency of the boiler when working at a definite pressure

   (c) to prepare heat balance sheet for the boiler

   (d) all of the above

   (e) none of the above

92. In a boiler, the heat is lost

   (a) to dry flue gases

   (b) in moisture present in the fuel

   (c) to steam formed by combustion of hydrogen per kg of fuel

   (d) all of the above

   (e) none of the above

93. A heat balance sheet for the boiler shows the complete account of heat supplied by 1 kg of dry fuel and heat consumed or lost.

   Agree ( )   Disagree ( )

94. The draught is the_______of pressures above and below the fire grate.

   (a) sum

   (b) difference

95. The object of producing draught in a boiler is

   (a) to provide an adequate supply of air for the fuel combustion

   (b) to exhaust the gases of combustion from the combustion chamber

   (c) to discharge the gases of combustion to the atmosphere through the chimney

   (d) all of the above

   (e) none of the above

96. The air pressure at the fuel bed is reduced below that of atmosphere by means of a fan placed at or near the bottom of the chimney to produce a draught. Such a draught is called

    (a) natural draught   (b) induced draught

    (c) forced draught   (d) balanced draught

97. The draught may be produced by a

    (a) mechanical fan   (b) chimney

    (c) a steam jet   (d) all of the above

98. A draught produced by a chimney due to the difference of densities between the hot gases inside the chimney and cold atmospheric air outside it, is called natural draught.

    Yes ( )   No ( )

99. The draught in locomotive boilers is produced by a

    (a) chimney   (b) centrifugal fan

    (c) steam jet   (d) none of the above

100.The draught produced by a steam jet issuing from a nozzle placed in the chimney is called

    (a) induced steam jet draught

    (b) chimney draught

    (c) forced steam jet draught

    (d) none of the above

101.The draught produced by a steam jet issuing from a nozzle placed in the ash pit under the fire grate of the furnace is called

    (a) induced steam jet draught (b) chimney draught

    (c) forced steam jet draught   (d) none of the above

102.The chimney draught varies with

    (a) climatic conditions   (b) temperature of furnace gases

    (c) height of chimney   (d) all of the above

    (e) none of the above

103.The mechanical draught produces more draught than natural draught.

    Agree( )   Disagree( )

104.The mechanical draught_______ the amount of smoke.

    (a) increases   (b) decreases   (c) does not effect

105.The efficiency of the plant ______with the mechanical draught.

    (a) increases   (b) decreases   (c) remains constant

106.Which of the following statement is wrong?

(a) The mechanical draught reduces the height of chimney.

(b) The natural draught reduces the fuel consumption.

(c) A balanced draught is a combination of induced and forced draught.

(d) all of the above

107. In a double acting steam engine, the steam is admitted on one side of the piston and two working strokes are produced during each revolution of the crankshaft.

Yes ( )   No( )

108.All steam engines work on

(a) Zeroth law of thermodynamics

(b) first law of thermodynamics

(c) second law of thermodynamics

(d) all of the above

(e) none of the above

109.Combustion of fuel takes place ________the engine cylinder of a steam engine

(a) outside

(b) inside

110.When steam after doing work in the cylinder passes into a condenser, the engine is said to be a

(a) slow speed engine

(b) condensing steam engine

(c) non-condensing steam engine

(d) vertical steam engme

111.When the expansion of steam is carried out in a single cylinder and then exhausted into the atmosphere or a condenser, the engine is said to be a simple steam engine.

Correct ( )   Incorrect ( )

112.A single acting steam engine produces   power than that of double acting steam engine.

(a) equal

(b) half

(c) double

(d) four times

113.Steam engines are called external combustion engines.

Yes ( )   No ( )

114.Which of the following statement is correct?

(a) A horizontal steam engine requires less floor area than a vertical steam engine.

(b) The steam pressure in the cylinder is not allowed to fall below the atmospheric pressure.

(c) The compound Steam engines are generally non-condensing steam engines.

(d) All of the above   (e) none of the above

115.The function of an eccentric is to provide motion to the side valve.

(a) Reciprocating   (b) rotary

REFERENCES

1. Parsons, Sir Charles A. "The Steam Turbine".

2. http://www.km.kongsberg.com/ks/web/nokbg0240.nsf/AllWeb/A297BDC3A79BBB36C125726B00387597?OpenDocument

3. http://www.guinnessworldrecords.com/world-records/431420-most-efficient-combined-cycle-power-plant

4. Horsley, D. Process Plant Commissioning, a User Guide, Institution of Chemical Engineering, 1998.

5. http://energy-models.com/boilers

6. http://www.respaper.com/bee_energy/563/2537-pdf.html

7. https://en.wikipedia.org/wiki/Combustion

8. https://en.wikipedia.org/wiki/Electricit

9. http://www.slideshare.net/hariomnagar1/k

10. http://textofvideo.nptel.iitm.ac.in/1121

11. http://www.readbag.com/seas-upenn-lior-l

12. http://www.google.it/patents/WO201212370

13. http://twt.mpei.ac.ru/TTHB/2/KiSyShe/eng

14. https://www.myodesie.com/wiki/index/retu

15. http://highered.mheducation.com/sites/dl

16. http://www.sciencedirect.com/science/art

17. http://www.sciencedirect.com/science/art

18. https://en.wikibooks.org/wiki/Engineerin

19. https://sseaimes.files.wordpress.com/201

20. http://www.sciencedirect.com/science/art

21. http://www.acr-news.com/masterclass-expa

22. http://www.ignou.ac.in/upload/Unit%205-3

23. http://www.slideshare.net/SatheeshSekar/

24. http://www.gutenberg.org/files/17921/179

25. http://www.erc.uct.ac.za/sites/default/f

26. http://docslide.us/engineering/power-pla

27. http://mhi-inc.com/Converter/watt_calcul

28. http://www.slideshare.net/waleedmahmoud1

29. https://www.scribd.com/document/65009227

30. http://textofvideo.nptel.iitm.ac.in/1121

31. https://www.researchgate.net/profile/Ian

32. http://twt.mpei.ac.ru/TTHB/2/KiSyShe/eng

33. https://www.scribd.com/doc/185638083/The

34. https://www.scribd.com/doc/47216966/250M

35. http://maritime.org/doc/fleetsub/still/c

36. http://documents.tips/documents/thermal-

37. https://www.scribd.com/doc/20285100/AIR-

38. http://www.em-ea.org/Guide%20Books/book-

39. http://journals.tubitak.gov.tr/elektrik/

40. https://docs.com/vinod-kumar/6113/guru-n

41. https://wasteadvantagemag.com/feasibilit